SpringerBriefs in Applied Sciences and Technology

Safety Management

The *SpringerBriefs in Safety Management* present cutting-edge research results on the management of technological risks and decision-making in high-stakes settings.

Decision-making in high-hazard environments is often affected by uncertainty and ambiguity; it is characterized by trade-offs between multiple, competing objectives. Managers and regulators need conceptual tools to help them develop risk management strategies, establish appropriate compromises and justify their decisions in such ambiguous settings. This series weaves together insights from multiple scientific disciplines that shed light on these problems, including organization studies, psychology, sociology, economics, law and engineering. It explores novel topics related to safety management, anticipating operational challenges in high-hazard industries and the societal concerns associated with these activities.

These publications are by and for academics and practitioners (industry, regulators) in safety management and risk research. Relevant industry sectors include nuclear, offshore oil and gas, chemicals processing, aviation, railways, construction and healthcare. Some emphasis is placed on explaining concepts to a non-specialized audience, and the shorter format ensures a concentrated approach to the topics treated.

The *SpringerBriefs in Safety Management* series is coordinated by the Foundation for an Industrial Safety Culture (FonCSI), a public-interest research foundation based in Toulouse, France. The FonCSI funds research on industrial safety and the management of technological risks, identifies and highlights new ideas and innovative practices, and disseminates research results to all interested parties. For more information: https://www.foncsi.org/

Stéphanie Tillement · Caroline Kamaté ·
Hervé Laroche · Corinne Bieder
Editors

The Shaken Foundations of Safety

Overcoming Human Resources and Skills Shortages

Springer

Editors
Stéphanie Tillement
IMT Atlantique
Nantes, France

Caroline Kamaté
FonCSI
Toulouse, France

Hervé Laroche
FonCSI
Toulouse, France

Corinne Bieder
FonCSI
Toulouse, France

ISSN 2191-530X ISSN 2191-5318 (electronic)
SpringerBriefs in Applied Sciences and Technology
ISSN 2520-8004 ISSN 2520-8012 (electronic)
SpringerBriefs in Safety Management
ISBN 978-3-032-22923-6 ISBN 978-3-032-22924-3 (eBook)
https://doi.org/10.1007/978-3-032-22924-3

This work was supported by Fondation pour une Culture de Sécurité Industrielle.

This Springer imprint is published by the registered company Springer Nature Switzerland AG
The registered company address is: Gewerbestrasse 11, 6330 Cham, Switzerland

Preface

The technological, socio-demographical, economic, and environmental megatrends affecting our world today and likely to intensify in the near future are at the core of FonCSI's reflections. As the Foundation's recent publications demonstrate, these evolutions are disrupting the risk landscape and impacting several dimensions of safety management and governance. This book, building on the previous work, focuses on the main difficulties related to human resources and the "skills challenge" that high-risk industrial sectors are and will be facing. These factors threaten the performance of organizations in multiple areas, and safety, whether it concerns facilities, employees, or the environment, is obviously at stake.

This publication is one of the outcomes of a strategic analysis, a working methodology developed by FonCSI, involving scholars as well as experts from its partner organizations. FonCSI therefore warmly thanks the architects of this analysis: Stéphanie Tillement (IMT-Atlantique, RESOH Chair, LEMNA), Florence Reuzeau and Lucille Guichardon (Airbus), Adeline Clos, Alexandre Largier, Audrey Marquet and Sophie Quiblier (ASNR), Stéphane Corcos (DGAC-DSAC), Raluca Ciobanu and Stéphanie Michaud (EDF), Romuald Périnet (NaTran), Alain Giot (SNCF), and Michel-André Klein (TotalEnergies). This book brings together the contributions of international experts invited to participate in a 2-day residential seminar held in November 2025, as well as those of some members of the working group.

What are the problems and lesser-known issues related to skills and career paths within industrial organizations? What avenues can be explored to address skills shortages through training? Through management? Here are the guiding threads of this three-part book, which never loses sight of the fact that professional skills and industrial safety are inextricably linked. Safety is part of the job.

Toulouse, France
January 2026

Caroline Kamaté
FonCSI

Contents

Chapter 1
Bridging Skills and Safety

An Introduction

Hervé Laroche

Abstract This opening chapter lays out the framework within which FonCSI's work has been situated for several years: safety in high-risk industries is inextricably linked to the professional skills of the individuals and groups working within organizations. It then expands on the implications of this assumption when, as this book does, one considers the broader skills challenges most industries already face, and which are expected to intensify in the near future. The final part of this introduction is a concise overview of the chapters that comprise the book.

Keywords Professional skills · Industrial safety · Staff shortages · Skills shortages · Training · Human resources · Organizational resilience

1.1 Industrial Safety and Professional Skills (or the Lack Thereof)

It is a fundamental assumption of this book that industrial safety is highly dependent on the people doing the job, regardless of the technology. Technology matters, of course, but as far as technical systems are designed, operated, monitored, maintained, modified, and more generally managed by humans in organizational contexts, the availability, quality, and commitment of the human resources devoted to these tasks are crucial factors in attaining the desired or required levels of safety. We must insist: this assumption extends much further than standard requirements for safety training, however developed or sophisticated this training may be (which, we all know, is not so common). Put simply: safety is part of the job; it is not something added to the job. This was the main conclusion of previous FonCSI work published in a book entitled *Beyond Safety Training* (Bieder et al. 2017). It follows that industrial safety and professional skills are inextricably linked. We understand professional skills here

H. Laroche (✉)
FonCSI, Toulouse, France
e-mail: herve.laroche@foncsi.org

Emeritus Professor, ESCP Business School, Paris, France

S. Tillement et al. (eds.), *The Shaken Foundations of Safety*,
SpringerBriefs in Safety Management, https://doi.org/10.1007/978-3-032-22924-3_1

in a very broad sense: all skills that are useful to do the job; not just the specialized, technical skills for operating the process or the machines.

The implication of this assumption is that, whenever some of these professional skills are lacking, safety is at risk. Professional skills may be lacking because people do not possess the skills (qualitative issue), because people themselves are lacking (quantitative issue), or both. Unfortunately, there are strong reasons to believe that both issues will develop in the years to come and, in fact, are already manifest in many industries. Because of a number of well-documented trends (unfavorable demographics, lack of attractiveness of industrial sectors, transformation of jobs in line with technological developments, digitalization, etc.), the industrial sector will have to deal with major difficulties concerning its human resources. High-hazard industries will be no exception (for more details, see FonCSI's report *Industrial safety in the age of "living with"* (Bieder et al. 2024) and its book *Managing Future Challenges for Safety* (Laroche et al. 2022)). In particular, and this will vary across industries and across countries, we can expect a lasting shortage of operational staff, a lack of professional skills or mismatches between the skills available and those required by the organization (Garcias et al. 2015). The issue of aging employees will also bring adaptation and transmission problems. Finally, the aspirations of new generations in relation to work will introduce new issues in the management of career paths (Gagné et al. 2022).

Of course, human resource shortages are nothing new. The recent COVID crisis provided some clues as to the adaptation required (Minbaeva and Navrbjerg 2023). Yet, what we are pointing to here is a structural issue of a greater magnitude. These difficulties threaten all aspects of company performance: the technical performance of products and services, the economic performance of financial results, and the social performance in terms of employee well-being. Safety performance, whether it concerns facilities, the environment or employees, is similarly at risk.

As always, technological developments are widely seen as promising solutions to these issues. And, certainly, some of them will help to tackle some problems. But, as we have witnessed in the past, other technological developments will only make these issues more severe, if only because their success will require new skills. And most of them will have both positive and negative effects, the balance of which is impossible to predict, even if this would make sense. Whatever one might think of the current AI craze, it is more than reasonable to expect that this technology will provide a way out of some of the skills shortages, sometimes by suddenly making a whole range of professional skills obsolete. Yet, it is also reasonable to expect that wide swathes of workers will have to adapt to AI-induced transformations in the nature of their tasks, with major problems ensuing in terms of work organization, training, and change management. Eventually, AI systems will reveal their own reliability weaknesses and we will need to develop the resources and the skills to address them. As it is not uncommon with sociotechnical systems, solutions come with new problems. There is no reason that digitalization, AI, and more generally the coming technological developments should not do the same with human resource challenges in the short and medium terms.

1.2 A Way of Reasoning about Professional Skills

We are well aware of the magnitude of the problem we address in this book, and of the complex, elusive, highly uncertain nature of the causalities we are trying to trace. Sweeping changes at one end, with various social and economic implications, and safety issues at the other end, in a variety of industrial and social contexts: it is no surprise that we could not rely on a unified, reliable model to develop our analyses. It would seem preposterous to even claim that we were able to identify the key variables and assess the most potent effects. This book does not pretend to provide a detailed mapping of the topic, nor even a stylized sketch. Rather, we suggest a way of reasoning, based on the following premises.

Firstly, we focus our attention on the company level or below, rather than adopting a macro perspective. Certainly, the macro level matters, as does the industry level, and we will occasionally touch on the corresponding institutions and policies. But, apart from initial skills, most decisions and policies relative to human resource and skills development are taken at the company level and are implemented (or not) at the shopfloor level. This is all the more relevant when we consider that our main interest is the implications of these decisions and policies for safety. We also rely a lot on examples that allow us to account for the complexities of the topic without oversimplifying.

Secondly, we adopt a dynamic view of professional skills, where time is a key dimension and, more than that, a causal variable in itself. It is tempting (and certainly legitimate) to view human resources as a pool of people and skills that companies should manage by keeping an exact log of what goes in and what goes out. This is what companies try to do. But, as we will see, they are unable to keep an exact log, and even less so to balance the inputs and outputs. We consider skills as having, to some extent, a life of their own. They develop, but they also wither away. They are nurtured by work as well as they nurture work. Sometimes they are explicitly defined, precisely located, and sometimes they are elusive and distributed (and often they are both at once). They may mutate, become obsolete, or be revived.

Thirdly, we keep in mind the social dimension of skills. Skills are owned by individuals but exercised in a collective context. They weave together into collective competencies or unweave themselves when a thread is missing. They are developed and exercised by people who have their own objectives, follow their own paths, leave too soon, or stay too long (from a company standpoint). These people also have their own idea of the skills they possess (or not), of how these skills should be used and maintained. Most of the time, skills are the basis for the development of social groups within and across companies, sometimes organized into professions. Skills are both a strategic asset and a key issue for these groups.

Within this triple framework, the skills problem appears as full of complexities, possible contradictions, and ambiguities. The first part of the book (*The conundrums of skills in organizations*) attempts to isolate and shed light on these mysteries. The second and third parts examine a range of solutions, or, to put it more humbly, explore two different (but complementary) ways of addressing skills shortages. Training obviously comes to mind when discussing skills. Training for what exactly? And how? Revisiting training from the perspective of a long-term shortage will be the subject of the second part (*Training solutions*). But new ideas and practices for training will not be enough to address the coming challenges. In the third part (*Managerial solutions*), the book suggests that skills management itself has to change and develop new approaches and new tools.

The final chapter offers a strategic view at the company level, by outlining three (analytically) distinct responses to skills shortage issues.

1.3 Book Outline

1.3.1 Part I—The Conundrums of Skills in Organizations

The first part of the book aims to unpack the main ambiguities and complexities that surround the topic, either by clarifying and discussing the key concepts or through case studies and examples.

The triple framework presented above needs more precise definitions. This is what Alexandre Largier and Audrey Marquet aim to provide in Chap. 2, "A Longitudinal View of Competencies: Careers, Professional Paths, and Trajectories." After retracing the development of the competency approach in Human Resource Management (HRM), they underline that competencies are intrinsically linked to the workers' autonomy and capacity for initiative. This is why companies try to manage them. Unfortunately, this is not easy. HRM systems tend to see them as knowledge accumulated by individuals and attempt to formalize them in normative frameworks (lists of competencies corresponding to lists of occupations). Competencies are still far more complex objects. Firstly, they include a collective dimension that HRM systems too often ignore. Secondly, they are constructed and evolve throughout the workers' professional paths, whether within a single organization or across multiple organizations. Here again, HRM practices have trouble acknowledging the acquisition of competencies in their career systems.

Note: Largier and Marquet use the word "competency" rather than "skill," because they see the latter as a limited notion, covering only individual abilities. In part, this preference for "competency" is the outcome of a heated debate that took place in France among HRM scholars, sociologists, and ergonomists. It appears that, outside France, the word "skill" is much more commonly used and covers a wider range of meanings, including the collective and dynamic dimensions underlined by Largier and Marquet. In subsequent chapters, the reader will find both words, with a strong preference for "skill." We invite the reader to consider both terms as synonyms.

In Chap. 3, "Managing Skills in a World of Radical Unpredictability: The Pioneering Experience of the Nuclear Industry," Frédéric Garcias examines the striking case of the French nuclear industry at the beginning of this century, when the construction of new reactors was resumed after being interrupted for many years. Unprepared and for a long time unaware of the emerging issue, the French Nuclear Operator faced a "learning wall" when the demand for training new employees exceeded the capacity of experienced staff to transfer their skills, resulting in serious project delays and difficulties. Analyzing how this sudden and massive shortage of skills was addressed by the French Nuclear Operator and the French government provides valuable insights for other sectors facing similar challenges in a world of increasing unpredictability.

In Chap. 4, "Digital Transitions in Industrial Work: Between Skills Development and Deskilling Risks," Marta Santos explores how digital transitions impact skills development, work autonomy, and collective dynamics. Drawing on two case studies from logistics operations within a multinational company, the analysis highlights opposite effects of technological change: in one case, it fosters learning and collaboration, and in the other, it leads to increased control, fragmentation, and deskilling. It reminds us that technological change has no univocal impact; it all depends on the context. This is why careful attention must be given by management to the implications of increasingly digitalized work environments for competence development and safety.

1.3.2 Part II—Training Solutions for Skills Shortages

Training being the most obvious answer when adequate skills are lacking, the second part explores this avenue. Still, as noted above, safety is part of the job in any context. Training should therefore go beyond standard, regulatory-imposed, safety-focused forms of training (Bieder et al. 2017). The three chapters composing this section concentrate on emerging practices, if not experiments in professional training. Embedded in specific professional contexts, these practices should not be taken as recipes that could be transposed in any industry or organization. Rather, they are presented as sources of inspiration because of the principles they rely upon to tackle the skills shortage issue, within the triple (organizational, dynamic, and social) framework adopted here.

In "Applying Resilience Engineering to Real-World Work" (Chap. 5), Elizabeth Lay draws on her experience of helping line clearance tree trimming teams to increase the reliability and resilience of individuals, teams, and systems through different types of training sessions inspired by Resilience Engineering principles. One key issue that this training has to address is the high turnover in these teams, which implies that less-skilled people enter the workforce at a high frequency. Lay proposes that, in such a context, training should be based on three key strategies. Firstly, it should be embedded in daily work practices. For instance, close calls should be systematically taken as opportunities for learning by sharing stories and experiences. Secondly, training should enable people to make safe, informed adjustments in response to real-world conditions, instead of insisting on compliance with rules that may not always fit. This "guided adaptability" strategy is a response to high variability at work. Thirdly, training should embrace ambiguity and uncertainty rather than trying to reduce or contain it. Noticing and revealing uncertainty should be encouraged and turned into opportunities for discussing adaptive change in work practices. One key inspiration that can be taken from this chapter is rethinking the articulation of training and working, in terms of both time and content.

Resilience training is also the focus of Chap. 6, "Tackling the "Eternal Beginners" Problem in Safety Management: The Case of Occasional Crisis Responders," but in a very different context. Elleke Ketelaars, Martin Viktorelius, and Simon Flandin argue that skills shortages increase the need for crisis preparedness in large sociotechnical systems. Still, conventional training approaches are focused on anticipated events, plans, and procedures. They produce standardized competencies but are unable to prepare people for handling complex, unanticipated situations. The authors suggest shifting the focus of training from general skills to situated resilience skills (embedded in the real work context) and developing the latter by introducing specific resilience-building exercises. Whereas usual exercises are centered on procedural knowledge, that is, on what is predictable, resilience-building exercises should introduce perturbations, disruptions, and indetermination. The collective dimension of skills should also be a focus of training. The authors argue that, when it comes to crisis management, we are all "eternal beginners," because we have to face the unknown. This should be a central tenet of training for crisis preparedness.

One extreme case of the unknown is the long-term need for technical skills. How do you build competencies for future tasks when you have only a vague idea of the technologies that will be available and of the practices that will be used? In Chap. 7, "Overcoming the Paradox of Training Design for Future and Ill-Defined Practices: The Case of Nuclear Dismantling Teleoperators," Laurane De Carvalho, Raluca Ciobanu, and Germain Poizat explore this challenge in the context of the dismantling of nuclear power plants in France. To date, very little is known about the tasks involved, except that this has never been done and that the tasks will be teleoperated. Designing training programs for these ill-defined practices requires not only the adoption of an anticipation stance but also the identification and selection of existing practices that will serve as a basis for preparing operators for their future jobs. Besides the nuclear industry, this practice approximation approach is presented here as an example potentially relevant for other industries that have a strong need for long-term preparedness.

1.3.3 Part III—Managerial Solutions for Skills Shortages

In the last part of the book, we try to widen the picture. If human resource and skills shortages are a lasting and profound challenge, then this is, in management parlance, a strategic issue that calls for top management attention and action. However much effort and imagination we invest in developing new forms of training, it will not be enough. This section outlines three different but complementary perspectives for inspiring and guiding managerial action in organizations facing human resource shortages.

Chapter 8, "Skills and Careers by 2040: Emerging Challenges and Skills-Based Solutions to Foster Employability of Workers," centers around the concept of sustainable employability. Developing and maintaining sustainable employability for workers is presented by Annet de Lange as a way of limiting and managing skills shortages. Sustainable employability might be seen from different angles. Workers themselves should be encouraged to keep their employability in mind and make relevant career choices. However, they only have limited power in this matter. They should benefit from an environment that helps them make the right choices and provides them with the resources to develop their skills. Companies are expected to design HR policies that not only improve individual career outcomes but also contribute to the sustainability of the labor market as a whole. In other words, they must ally themselves with a variety of actors and stakeholders to develop ecosystems that favor skill development. Annet de Lange exposes the features of such ecosystems and the steps needed to develop them.

In Chap. 9, Manuel Souto-Otero conducts "A Review of Organizational Strategies" on "How to Address Skills Gaps". Obviously, companies may seek to acquire the needed skills (capability expansion) by recruiting, training, or contracting. They may also encourage a more efficient use of available skills (insourcing), especially by stimulating an internal market for talent. Finally, they may consider adapting their product and production processes to align with available skills (redesign). Redefining or reducing the skills content of jobs is rarely considered, because it is not a standard approach in Human Resources Management. All these strategies, however, depend on an accurate appraisal of skills gaps. This, as Souto-Otero explains, is a skill that few organizations possess.

Chapter 10 explores the "Competencies and Skills for Navigating Uncertain and Interdependent Systems of the Future." Building on the concept of dynamic capability, Mark A. Griffin outlines a framework for encompassing both individual and organizational levels. Dynamic capabilities cover a company's ability to sense changes in the environment, seize opportunities, and transform itself accordingly. In other words, dynamic capabilities support strategic adaptation in the face of uncertainty. A key feature of this chapter is that strategic adaptation is viewed as a process involving opening and closing cycles: opening to uncertainty and fostering change, or alternatively removing uncertainty and fostering stability. Focusing on safety, the chapter closes on an analysis of what constitutes a dynamic safety capability and on how it might be managed.

The concluding chapter "Meeting the Skills Challenge: Fostering a Compensation Strategy for Safe Performance" seeks to deepen the previous analyses through the lens of the strategies that organizations may deploy to address the challenge of maintaining and developing skills under conditions of uncertainty and pervasive scarcity. Firstly, companies may do nothing or not much, simply ignoring the issue or counting on the virtues of local adaptation and resilience. This "adjustment" response implies a tacit acceptance of degrading working conditions and elevated risks, including a possible "normalization of misery." Secondly, companies, recognizing the challenge, may increase their efforts to return to a predefined reference level of resources and competencies by designing and implementing a "restoration" strategy, relying on traditional HRM approaches: forecasting, workforce planning, and standardized training. While often necessary, this "more of the same" approach appears poorly suited to situations of acute or rapidly evolving scarcity. Thirdly and finally, companies may acknowledge the limitations of the restoration path and engage in a "compensation" strategy. Compensation consists in adopting a logic of organizational reconfiguration, driven by a redistribution of skills where they are most needed to perform safely. Compensation implies rethinking how work is organized in order to enhance the effective use of existing competencies. Implementing such a strategy requires profound organizational change and challenges established distributions of authority and responsibility, which points to a central question: who assumes responsibility for organizing and equipping skills maintenance and development so as to sustain safe performance? The respective roles of top management, HRM actors, operational staff, and safety managers have to be redefined and articulated to support the organizational structures and processes necessary to the compensation strategy.

References

C. Bieder, R. Amalberti, J. Pariès, H. Laroche, E. Marsden, C. Kamaté, *Industrial Safety in the Age of "Living with": Uncertainty, Complexity and Rising Expectations* (FonCSI: Foundation for an Industrial Safety Culture, Toulouse, 2024) https://www.foncsi.org/sites/default/files/Publications/Publications_EN/CSI_2024-02_safety-living-with.pdf

C. Bieder, C. Gilbert, B. Journé, H. Laroche, *Beyond Safety Training: Embedding Safety in Professional Skills* (Springer, Cham, 2017). https://doi.org/10.1007/978-3-319-32939-0

M. Gagné, S.K. Parker, M.A. Griffin, P.D. Dunlop, C. Knight, F.E. Klonek, X. Parent-Rocheleau, Understanding and shaping the future of work with self-determination theory. Nat. Rev. Psychol. **1**(7), 378–392 (2022)

F. Garcias, C. Dalmasso, J.C. Sardas, Paradoxical tensions in learning processes: Exploration, exploitation and exploitative learning. M@n@gement **18**(2), 156–178 (2015)

H. Laroche, C. Bieder, J. Villena-Lopez, *Managing future challenges for safety -demographic change, digitalisation and complexity in the 2030s* (Springer, Cham, 2022). https://doi.org/10.1007/978-3-031-07805-7

D.B. Minbaeva, S.E. Navrbjerg, Strategic human resource management in the context of environmental crises: A COVID-19 test. Hum. Resour. Manag. **62**(6), 811–832 (2023)

Part I
The Conundrums of Skills in Organizations

Chapter 2
A Longitudinal View of Competencies

Careers, Professional Paths, and Trajectories

Alexandre Largier and Audrey Marquet

Abstract Competencies are developed and acquired within work situations but also unfold over the course of professional life when considered from a longitudinal perspective. Competencies are not merely individual abilities; they also have an important collective dimension. They are dynamic, situated, and distributed, linked to the technical, social, and organizational environment. At the end of the post-World War II economic boom, companies seeking greater flexibility looked to draw on their employees' capacity for initiative, leading to the emergence of competency management in the 1990s. Competency management promoted the individualization of professional paths, marking a departure from career models in which progression was largely tied to seniority. However, careers and professional paths often coexist. Finally, it is worth considering inter-organizational trajectories, which are often facilitated by digital technologies and also shape competencies. Activities carried out outside the company enable individuals to develop new competencies. This raises the issue of how employers can integrate and recognize these.

Keywords Competencies · Careers · Professional paths · Trajectories

2.1 Introduction

Although the question of competencies is undeniably linked to the work situations in which they are applied, it is also necessary to consider how they are acquired and developed over the long term. Broadly speaking, this acquisition can take place through careers, paths, and trajectories in which individuals acquire and develop competencies during daily work situations or in spaces dedicated to learning, whether within or outside the company.

Before discussing the concepts of careers, paths, and trajectories, it is important to define the term competencies. Ergonomics and sociology provide complementary approaches, each providing a distinct perspective on the processes through which individuals develop as professionals.

A. Largier (✉) · A. Marquet
ASNR (The French Authority for Nuclear Safety and Radiation Protection), Paris, France
e-mail: alexandre.largier@asnr.fr

S. Tillement et al. (eds.), *The Shaken Foundations of Safety*,
SpringerBriefs in Safety Management, https://doi.org/10.1007/978-3-032-22924-3_2

2.2 Defining Competencies

The theme of competencies first emerged in company discourse and union demands in the early 1970s. It marked the decline of Fordist regulation (Paradeise and Lichtenberger 2001, p. 36), which had provided stability in labor relations after World War II. By the end of three decades of economic development and growing prosperity (a period known in France as the “Trente Glorieuses”), companies seeking greater flexibility looked to harness their employees’ capacity for initiative. The development of the concept of competency was therefore linked to the need to give greater consideration to contingencies, uncertainties, novelties, and anomalies—all destabilizing factors requiring vigilance, responsiveness, and initiative.

In ergonomics, Montmollin (1997) offers a widely cited definition describing competencies as stabilized sets of knowledge, know-how, reasoning, and typical behaviors that can be mobilized without new learning and that are embedded in the individual’s professional history. These elements enable an individual to anticipate, adapt to task variability, and understand instructions implicitly. It is also important to emphasize that competencies cannot be reduced to individual abilities (Montmollin 2001; Leplat 2000). Caroly and Cholez (2007) remind us that competencies are also shaped by the collective contexts in which they are applied. They develop through social interactions among colleagues or with management and are often informally shared among members of the same work group.

In sociology, the focus is more on the social and organizational implications of competency. Reynaud (2001) distinguishes competency from qualification[1] by emphasizing that the former implies greater employee responsibility: the worker is no longer merely an executor but also responsible for the outcomes of their actions. This definition introduces a subjective and relational dimension, linked to the autonomy expected of the actor. According to Paradeise and Lichtenberger (2001), competency is manifested in the ability to effectively mobilize not only one’s own resources but also those available within the networks of interdependence that characterize work groups. Competency thus becomes a strategic resource, used in adaptive approaches aimed at problem-solving and adjusting to unforeseen events.

Although they differ, these two approaches converge in that they view competency as a situated, dynamic and contextual process far removed from a mere accumulation of knowledge. It entails an active and reflective relationship to work, influenced by technical, social, and organizational environments.

Finally, the professionalization and development of workers’ competencies cannot be analyzed independently of the regulatory and institutional frameworks in which they are embedded. Standards, laws, and directives, whether national or international, help structure the processes for training, evaluating, recognizing, and certifying competencies by defining both what is expected of professionals and the conditions under which those competencies must be acquired and mobilized. Public policies

[1] Qualifications are linked to the job position and define the credentials (degrees, certifications, accreditations, etc.) required, as well as the level of seniority necessary to hold that position (Dubar 1996). In contrast, competencies are linked to the individual.

and regulatory frameworks contribute to the formalization of professional competencies, particularly in regulated sectors (healthcare, education, nuclear, etc.). These normative frameworks aim to make the knowledge required for professional practice explicit and standardized (Barbier 2011), addressing managerial objectives such as safety, quality of service, and professional mobility (Lefresne 2007). However, this standardization can clash with the reality of work situations. Real competencies are constructed in and through activity (Leplat 2000), in unique and variable contexts, yet competency management systems often fail to capture work-as-done (Teiger and Daniellou 2007). First, they reduce the number of expected competencies in management tools. Second, they tend to individualize competencies, whereas the collective dimension of activity is crucial. Finally, they often provide a static view, whereas competencies are a living, evolving matter, shaped by lived experiences and the positions held.

Competencies, as defined in their individual, collective, and institutional dimensions, are a central element of professional activity. They are constructed and evolve over time, in interaction with work contexts. This is why they must be considered within the longitudinal perspective of careers and professional paths.

2.3 Careers and Paths

Due to their dynamic and situational nature, the acquisition and development of competencies are necessarily long-term processes that take place within the context of careers or professional paths. It is therefore useful to clarify what is meant by "path" and "career." A career, as conceptualized by Hughes (1937) and Becker (1985), refers to the orderly sequence of movements from one position to another within a given system. It is understood as a succession of predetermined stages, most often planned by company management, with rules and structures defined by the organization. According to Zimmermann (2011), a career is a form of professional advancement officially recognized by a competent authority, such as a company, which validates progress through changes in position, level of responsibility, or remuneration. This advancement takes place within a structured organizational framework, where access to the different positions is based on a logic specific to that environment.

Furthermore, individuals follow two types of paths: professionalization paths (Marquet 2021) and professional paths (Monchatre 2007).

In the context of a career, individuals develop their competencies throughout their professionalization path (Weill-Fassina 2012) based on a host of experiences influenced both by everyday work situations and those designed to promote learning (mentoring, initial training, apprenticeship-style guidance, etc.). In this sense, there is a diachronic dimension (past, present, future) to competency development, articulating individual, collective, and organizational timeframes and spaces of production and training, which overlap, intertwine, and reconfigure themselves throughout the career (Delgoulet 2015).

In theory, with competency management in companies, careers are supposed to give way to professional paths—the latter expression suggesting a progression "free" from constraints (Monchatre 2007). The "competency approach" (Zarifian 1988, 1999; Durand 2000) emphasizes individual recognition of performance and therefore promises differentiated professional paths based on each person's investment and quality of work. Competency management puts an end to automatic career progression based on seniority and "individualizes promotion decisions based on an assessment of each person's productive contributions" (Monchatre 2007, p. 515). The question of individual recognition of each person's contribution arises, especially as contemporary work is increasingly collaborative (project-based work, cross-functional teams, etc.). In practice, careers and professional paths most often coexist within organizations. Under the "competency approach," individuals are expected to take a proactive approach to managing their professionalization, yet a more directive conception of career progression is still at work. For example, professional or technical streams encourage certain employees to develop as professionals by holding a series of positions. Certain companies establish rules outlining standard professional paths through which employees acquire and develop competencies. This is particularly the case for professions or roles where technical mastery can only be achieved over a long period of time (certain engineering professions, for example; Largier et al. 2008).

For competency management to foster a virtuous process of professional recognition, particularly through promotion, the internal labor market must be dynamic and offer concrete opportunities for mobility (Monchatre 2007). However, this is only feasible in organizations where such mobility is possible and the internal market is sufficiently large. Otherwise, blockages can arise (Hatzfeld 2004) in the course of professional life, potentially leading to withdrawals or even departures.

2.4 Inter-organizational Trajectories and Competency Acquisition

These withdrawals and departures, whether in the form of resignations, extended leave (for personal reasons, training, business creation, etc.), or taking on a second professional role, are in a sense consistent with the "competency approach": it is the individuals themselves, professional employees, who assume partial responsibility for developing their competencies. However, the acquisition and development of new competencies occur outside the boundaries of the organization. This gives rise to professional paths that break with predetermined careers or individualized professional paths within an internal labor market.

The development of digital platforms and information and communication technologies (ICT) greatly facilitates the development of these trajectories, which transcend company boundaries and existing careers and professional paths. Some studies (Beauvisage et al. 2018) show, for example, that digital platforms and ICT are reshaping, and even facilitating, the development of activities that blur the line between work and leisure[2] (language training, volunteering, etc.). However, these "other activities" rarely replace the primary salaried activity. ICT enable a hybridization of salaried employment and entrepreneurship and the development of new working practices (Bohas et al. 2018). Thus, this "side activity" (Weber 2009) is most often conducted alongside the main professional activity. This hybridization, which forms the basis of the "slasher"[3] phenomenon (Alboher 2007; Bohas et al. 2018), provides an opportunity to expand knowledge and develop new competencies. In this respect, these individuals play an active role in, and are responsible for, the acquisition and development of their competencies. "Side" activities may also be motivated by a desire to perform one's job more efficiently. Some individuals with rare, specialized competencies or expertise in a particular field are sometimes tempted to put their skills to work for others in a setting they consider less restrictive. For example, some customer advisors participate in discussion forums outside of their working hours to continue informing users, but in a less prescriptive context (Klein 2016). While engaging in "side" activities does indeed fall under the competency approach, these activities largely occur outside the organization in which the employee is primarily employed. To what extent can these new competencies be mobilized in the primary activity? And even if these new competencies could be applied in future roles, the employer would first need to be aware of these activities and what they entail in order to integrate the ensuing competencies into a professional path.[4]

If these withdrawals and departures are not anticipated or supported, they can be detrimental to the company. This is particularly true when the acquisition and development of competencies is a long-term process and when the implementation of these competencies relies on the mobilization of other individuals and on competencies scattered across the organization (Largier et al. 2008). A resignation or involvement in a "side" activity can disrupt or weaken the long process of acquiring competencies in the context of a career or a professional path. That is why it is important for organizations and management to foster trajectories that enable, on the one hand, the acquisition of expert competencies, and on the other, the creation of an interpersonal network or the development of an organizational competency[5] (Largier 2023), in order to mobilize support and resources distributed throughout the organization and beyond. To this end, where relevant and possible, companies would benefit from considering inter-organizational trajectories in the acquisition and development of competencies, and from incorporating the possibility of temporary external mobility within careers and professional paths. Offering assignments,

---

[2] One notable consequence of the widespread adoption of ICT is the interpenetration of the private and professional spheres, accompanied by what some authors describe as a blurring of organizational boundaries and even a rethinking of these categories (Enlart and Charbonnier 2013).

[3] Named after the typographical symbol "/" that separates their different activities.

[4] This is also the case for competencies developed in the context of community, union, or leisure activities.

[5] Organizational competency refers both to the ability to adapt to the organization and to the ability to adapt the organization in order to act.

secondments, and visiting positions, with return clauses, can encourage the acquisition and development of competencies that benefit the organization, while preserving the employment relationship.

2.5 Conclusion

Ultimately, the acquisition and development of competencies cannot be considered independently of the social, organizational, and temporal contexts in which they unfold. While the "competency approach" promotes individualized paths and places greater responsibility for learning and development on the employee, it often clashes with the collective and contextual realities of work. The emergence of inter-organizational trajectories, enabled in part by digital technologies, challenges the existence of predetermined careers and structured professional paths within an internal job market, and invites us to rethink the boundaries of professional learning and the ways in which competencies are recognized. For mutual enrichment to occur, organizations may need to adapt their management practices to integrate these hybrid paths and to capitalize on experience gained externally.

In sectors such as the nuclear industry, where professional mobility already occurs within what some authors describe as a "small world" (Mangeon and Pallez 2017), supporting such mobility may be preferable to focusing exclusively on staff retention. Indeed, employees may value an organization that prioritizes their development, supporting them even when they desire to pursue opportunities outside the company, and keeping the door open for a future return.

This chapter opens a perspective on considering the longitudinal and inter-organizational dimensions of competencies, which in turn leads to the following questions: To what extent does the company encourage or discourage the recruitment of inter-organizational profiles? Is it able to identify, assess, and capitalize on competencies acquired beyond its walls? Does it support employees in their inter-organizational trajectories, and if so, in what ways?

References

M. Alboher, *One Person/Multiple Careers.* (PT Mizan Publika, 2007)

J.-M. Barbier, *Savoirs théoriques et savoirs d'action* (Presses Universitaires de France, Paris, 2011)

T. Beauvisage, J.-S. Beuscart, K. Mellet, Numérique et travail à-côté. Enquête exploratoire sur les travailleurs de l'économie collaborative. Sociol. Trav.**60**(2) (2018). https://doi.org/10.4000/sdt.1984

H. Becker, *Outsiders.* (Métailié, 1985)

A. Bohas, J. Fabbri, P. Laniray, F.-X. de Vaujany, Hybridations salariat-entrepreneuriat et nouvelles pratiques de travail : des slashers à l'entrepreneuriat-alterné. Technol. Innov.**18**(1) (2018)

S. Caroly, C. Cholez, Compétences individuelles et compétences collectives : regard croisé entre sociologie et ergonomie, in *Travail et reconnaissance des compétences*. ed. by W. Cavestro, C. Durieux, S. Monchatre (Économica, 2007), pp.15–30

C. Delgoulet, *L'expérience à l'épreuve des apprentissages professionnels : conflit ou harmonie ? Construire la pertinence interne et externe des dispositifs d'apprentissage pour le développement des femmes et des hommes au travail* [Habilitation à diriger des recherches, Université de Bordeaux] (2015)

C. Dubar, La sociologie du travail face à la qualification et à la compétence. Sociologie Du Travail **2**(38), 179–193 (1996)

J.-P. Durand, Les enjeux de la "logique compétence." Gérer Et Comprendre **62**, 16–24 (2000)

S. Enlart, O. Charbonnier, *À quoi ressemblera le travail demain ?* (Dunod, 2013)

N. Hatzfeld, L'individualisation des carrières à l'épreuve. Les grippages de la mobilité sur les chaines de Peugeot Sochaux. *Sociétés contemporaines* (**2**) n°54, 15–33 (2004)

E.C. Hughes, Institutional office and the person. Am. J. Sociol. **43**(3), 404–413 (1937)

N. Klein, La numérisation de la relation de service : quelle place pour la dimension commerciale ? Le cas des webconseillers d'Orange, in *L'organisation face aux transformations de la relation de service*. ed. by M. Benedetto-Meyer, J. Cihuelo, A. Largier (L'Harmattan, 2016), pp.27–55

A. Largier, C. Delgoulet, C. De la Garza, Quelle prise en compte des compétences collectives et distribuées dans la gestion des compétences professionnelles ? *Perspectives interdisciplinaires sur le travail et la santé*, (10–1) (2008). https://doi.org/10.4000/pistes.2167

A. Largier, *Les compétences de l'expertise. Le cas de la fabrication d'évaluations de sûreté nucléaire à l'IRSN* [Habilitation à diriger des recherches, Université Gustave Eiffel] (2023)

F. Lefresne, Formation tout au long de la vie : des enjeux convergents dans des contextes institutionnels différents au sein de l'Europe, in *Travail et reconnaissance des compétences*. ed. by W. Cavestro, C. Durieux, S. Monchatre (Économica, 2007), pp.97–114

J. Leplat, Compétences individuelles, compétences collectives. Psychol. Trav. Organ. **6**, 47–73 (2000)

M. Mangeon, F. Pallez, Réguler les risques nucléaires par la souplesse : genèse d'une singularité française (1960–1985). Annales des Mines - Gérer Et Comprendre **130**(4), 76–87 (2017)

A. Marquet, *Principes d'évaluation de la professionnalisation pour travailler en sécurité : le cas des métiers de la circulation d'une entreprise du ferroviaire* [Thèse de doctorat, Université Paris Cité] (2021)

S. Monchatre, Des carrières aux parcours… en passant par la compétence. Sociologie Du Travail **49**(4), 514–530 (2007)

M. Montmollin, *Les compétences en ergonomie* (Presses Universitaires de France, 2001)

M. Montmollin, *Sur le travail : Choix de textes (1967–1997)* (Octarès, 1997)

C. Paradeise, Y. Lichtenberger, Compétences, compétence. Sociologie Du Travail **43**(1), 33–48 (2001)

J.-D. Reynaud, Le management par les compétences : un essai d'analyse. Sociologie Du Travail **43**(1), 7–31 (2001)

C. Teiger, F. Daniellou, Intervenir sur le travail. In P. Falzon (Éd.), *Ergonomie*. (PUF, 2007), pp. 433–448

F. Weber, *Le travail à-côté*. Éditions de l'EHESS (2009)

A. Weill-Fassina, Le développement des compétences professionnelles au fil du temps, à l'épreuve des conditions de travail, in *La vie professionnelle : âge, expérience et santé à l'épreuve des conditions de travail*. ed. by C. Gaudart, A.-F. Molinié, V. Pueyo (Octarès, 2012), pp.117–144

P. Zarifian, L'émergence du modèle de la compétence. In F., Stankiewicz, *Les stratégies d'entreprise face aux ressources humaines*, p. 77–82. Économica (1988)

P. Zarifian, *Objectif compétence. Pour une nouvelle logique*. Éditions Liaisons (1999)

B. Zimmermann, *Ce que travailler veut dire. Une sociologie des capacités et des parcours professionnels*. Économica, coll. "Études Sociologiques" (2011)

Chapter 3
Managing Skills in a World of Radical Unpredictability

The Pioneering Experience of the Nuclear Industry

Frédéric Garcias

Abstract The text discusses the challenges faced by the nuclear industry in France regarding workforce development and skills management. It focuses on the phenomenon of the "learning wall," which occurs when the demand for training new employees exceeds the capacity of experienced staff to effectively transfer their skills. This imbalance creates tension between two competing needs: advancing projects and training future engineers, resulting in a detrimental impact on both workforce development and project progress. This situation was encountered by the actors of the French nuclear industry when the construction of new reactors was resumed after a hiatus of many years. The chapter explores how the French nuclear operator and the French government addressed these issues. It suggests that the experience of the nuclear industry could provide valuable lessons for other sectors facing similar challenges in a world of increasing unpredictability.

Keywords Learning wall · Skills · Training · Nuclear reactors development

3.1 Introduction

Having studied the development of new reactors in the nuclear industry for about 15 years, in this chapter I will explore a phenomenon that I have observed within this sector, which I have chosen to call the "*learning wall*." This expression refers to a sudden and large-scale intensification of the need to train individuals in specific skills, to the point that the development of competencies creates tension with the other objectives of organizations, generating unsolvable dilemmas for teams. Although the nuclear sector has its own specificities, including its capital intensity, political ties, and the prevalence of strict regulations, I am convinced that the mechanisms I will describe can apply to other industrial sectors. Due to the unique historical

F. Garcias (✉)
University of Lille, Lille, France
e-mail: frederic.garcias@univ-lille.fr

S. Tillement et al. (eds.), *The Shaken Foundations of Safety*,
SpringerBriefs in Safety Management, https://doi.org/10.1007/978-3-032-22924-3_3

deployment of its activities, the French nuclear industry is, in my opinion, one step ahead in the world of radical unpredictability that has characterized the contemporary era since 2020. In this "new world," anticipating resource needs (which arise from answering the following questions: what activities will exist tomorrow? what jobs will be needed? at what scale will we need which competencies?) becomes a highly perilous exercise, yet undoubtedly one of the most strategic tasks. Therefore, the experience gained by stakeholders in the French nuclear industry over the past two decades constitutes a kind of "feedback," that is, a valuable resource upon which other sectors and organizations confronted with similar issues may rely in the future. Through my collaborations with the French nuclear operator and other organizations in the sector, I have gained a glimpse of the issues encountered.

3.2 The French Nuclear Industry: A Brief Overview of the Historical Context

In terms of competency management, the French nuclear industry is particularly unique. The history of this industry, particularly as a result of the political decisions of the Messmer Plan in March 1974 (a decision made by the French Prime Minister following the first oil shock of late 1973), has shaped a unique situation (Garcias and Tillement 2024). During the 1970s and 1980s, the French nuclear operator and its partners (such as Framatome) pursued an intensive policy of developing new infrastructure, installing 58 reactors over two decades. This period allowed France to become a global leader in nuclear electricity production, contributing to the country's energy sovereignty and the reduction of electricity production costs, while also fostering a low-carbon footprint. The construction of the French nuclear fleet is often presented as a genuine industrial feat (Grubler 2010) and a kind of "energy transition" ahead of its time, as it reduced France's dependence on fossil sources like coal and oil. However, this period of rapid growth was followed by a marked slowdown starting in the mid-1980s, notably due to the end of the Messmer Plan, doubts about the future of nuclear energy, and a reduction in investment in new projects. The Messmer Plan was the first wave of construction, but it was not immediately followed by a second wave. Thus, starting in the mid-1990s, faced with the uncertain future of its nuclear business, the French nuclear operator and its partners saw a significant portion of their human capital erode, with many engineers leaving for other sectors or retiring without being replaced by new recruits. This disenchantment affected not only the company but also the entire industrial base of the sector, with suppliers of components and subcontractors being forced to reduce their workforce due to falling demand.

Starting in 2005, the nuclear industry seemed to experience a global "renaissance" (Grimes and Nuttall 2010), with the launch of new projects. The growth of Asia, particularly China, brought renewed hope to nuclear operators, and the French sector aimed to get back into the race. The launch of the first new-generation PWR *(Pressurized Water Reactor)* construction project in 2007 marked what seemed to be a turning point toward the long-awaited "second wave" of construction of new reactors. This project had two main objectives: first, to test a new technology, the Generation III reactor (which integrates the feedback from the Generation II pressurized water reactors while addressing the new risks of serious accidents), and second, to revive a network of skills and expertise that had become more or less "dormant" within the nuclear sector, which was suffering from an aging workforce and the loss of experience accumulated over decades. One of the critical challenges for the French nuclear operator was, notably, to use this project to give a large number of young engineers and technicians the opportunity to build the "second wave" of the French nuclear fleet, as the reactors built in the 1970s and 1980s were beginning to approach the end of their theoretical 40-year lifespan.

However, despite the stated ambitions, this project quickly suffered delays, defects, and significant cost overruns. Originally scheduled to last 5 years and be completed in 2012, it ultimately took an additional 12 years, with the reactor being commissioned in 2024. In the media, this became a quagmire, particularly for the French nuclear operator, with the press consistently highlighting the "exorbitant" delays (without ever specifying that the French nuclear operator's last reactors took over 10 years to build and that the initial 5-year schedule was mainly political and commercial signaling). These delays were often caused by modification requests issued by the safety authorities as a result of identified non-conformities or potential defects posing safety risks. The delays could result either from work stoppages required to carry out inspections or from modification works requested by the safety authorities.

3.3 The "Learning Wall": Causes and Consequences of an Insidious Mechanism

Although not unique to the nuclear industry, and recurring in what some authors have called "mega-projects" (Flyvbjerg 2008), this issue of delays and cost overruns in project implementation triggered turmoil up to the highest levels of government, to the point where the Minister of the Economy ordered a report on the subject in 2019 (report by Jean-Martin Folz). This report emphasizes a "massive loss of skills," although it does not elaborate on the mechanisms of the phenomenon. In my work, I sought to understand how a "learning wall" can form within an organization, drawing from long-standing research on organizational learning through practice and learning curves (Argote 1996; Argote and Epple 1990). This phenomenon arises when the demand for on-the-job training for new employees exceeds the ability of experienced employees to effectively transfer their know-how, thus creating an imbalance between "productive" work and "training" work.

One of the main causes of this phenomenon lies in the specialized nature of the required skills: in the nuclear industry, many of the necessary skills cannot be acquired directly from the labor market (Barney and Arikan 2005). They must be developed and transmitted within the organization itself, through mechanisms of mentoring during and through work activities. This implies that young recruits, although theoretically qualified through their studies, cannot be immediately "productive" (Van Oorschot et al. 2013). They must first be trained internally, which requires constant interaction with more experienced engineers. In this context, the work of skill building cannot be dissociated from "productive" work. In other words, the quality of skill transfer depends directly on the demographic structure of the teams and the working conditions, which either enable or hinder this transfer.

When experienced teams are too few compared to the large number of young recruits to train, a trade-off emerges between the need to advance projects and the need to train young engineers (Garcias et al. 2015). But how does one choose between advancing current projects and developing the skills essential for future ones? This dilemma resembles a form of "intertemporal tension" (Slawinski and Bansal 2015), or what financial and climate experts call the "tragedy of horizons" (Silvi & Rosa 2023): the feeling of having to choose between the present and the future. The consequences are significant: internal tensions emerge and the quality of skill transfer deteriorates.

One of the most insidious risks is the "learning trap," which makes the training of a new generation of engineers resemble a Sisyphean task. Indeed, the learning wall creates a tense, degraded situation, which generates the risk of increased staff turnover and difficulties in attracting new graduates. This in turn perpetuates the lack of skills and resources, and the organization never fully recovers. My research work has shown that the phenomenon of "organizational forgetting," which is defined in the literature as "the loss, voluntary or otherwise, of organizational knowledge" (Holan and Phillips 2004), is an issue more insidious than a mere temporary loss of know-how: it is a potentially self-reinforcing mechanism (or a "negative feedback loop"), in which the consequences reinforce the very conditions that created it.

Another characteristic of the "learning wall" studied is that it can remain relatively invisible or imperceptible to those within the organization. Unlike tangible elements such as the production of plans or technical specifications, skills management is a process that involves working with informal, relatively uncertain mechanisms which are usually only empirically understood and over which one can only have partial control: two individuals exposed to the same conditions for skill development will not necessarily progress at the same rate. Similarly, calling someone "competent" is a social construct. This is why tools aimed at formalizing skill management, giving it the appearance of "engineering," are always very imperfect, especially in the eyes of nuclear engineers accustomed to deterministic reasoning! This explains why, in the early years of my study of the French nuclear operator, which coincided with the first difficulties encountered on the new-generation PWR project, the issue of skills was rarely seen as a critical challenge for the organization. Often, other factors were considered a priority: improving project management methods to make workload forecasts more reliable, for example (Garcias et al. 2024).

Two main reasons explain why this loss of skills is often overlooked by those in the organization:

1. **The Intangibility of Human Capital and Learning Processes:** The mechanisms of skill transfer are often informal and invisible, making them difficult to

model. In an organization where performance is measured in terms of tangible technical results, skills, which are crucial for the organization's sustainability, are often underestimated and under-theorized.

2. **The Equifinality of Performance and Non-performance Causes** (Gelhard et al. 2016)**:** When the organization faces operational problems, several factors may be cited as explanations: technological innovations, increased regulatory requirements, or changes in the societal environment. In this context, the loss of skills is hard to identify among other competing factors.

3.4 Recognition and Management of the Skills Crisis

This difficulty in recognizing skill management as a key factor in explaining difficulties, and in then understanding the underlying mechanisms, risks wasting precious time for an organization engaged in a race against the clock to rebuild its human capital. Fortunately for the organization, this awareness eventually emerged in several stages. This was not so much due to our research, which did not receive significant attention from the French nuclear organizations at the time it was conducted. Things began to change with the Folz report in 2019, when the assessment was made externally by an authority viewed as legitimate by nuclear leaders (Jean-Martin Folz being a former industry executive), and at the government's request. This shift in the industry's stance on the issue of skills can also be explained as a shift in the sector's own interests: initially (notably at the time of my own research in 2011–2012), if publicized, the thesis of a "loss of skills" may have been viewed as potentially damaging to the competitiveness of French firms at a time when they were seeking international contracts. By 2019, and subsequently, the situation had changed for the French nuclear industry: it was now a matter of convincing the French government of the need to engage in a new reactor program, and the argument of a loss of skills due to an overly small and erratic number of state orders in recent decades could be seen as aligned with the industry's interests.

Measures were then taken to stop the "negative feedback loop" related to the "learning wall": identifying the most critical areas in terms of experienced resources and restoring balance, accepting longer tenures in key positions to facilitate knowledge capitalization, implementing proactive mentoring systems, encouraging new recruits to immerse themselves on-site, and finally revising the allocation of tasks between engineering and subcontracting to facilitate career paths enabling rapid skill development for young recruits. All engineering managers for new reactor teams were thus trained to understand the mechanisms of skill loss and rebuilding, and, importantly, the role a team leader can play in this process. A "*Université des Métiers du Nucléaire*" (UMN, a national institute for nuclear professions) was even created at the national level to improve coordination between stakeholders in the sector, particularly in terms of ties with educational institutions, but also to avoid potential conflicts over access to the most qualified resources.

A final major point: after several years of critical analysis of the skills issue, the French government recognized that an industrial skill base could only be built on a long-term perspective. This is why it decided to give the sector's companies more visibility by announcing not a single "project," like "new-generation PWR," but a series of reactors (a new "program"). This choice reflects a recognition within the industry that maintaining engineering skills is impossible without maintaining technological development activities, that is, without a continuous flow of projects that constantly updates the knowledge and know-how embedded within an industrial ecosystem. This idea ultimately echoes both the literature on organizational forgetting, which has long identified that "learning curves" can regress when activity slows down or stops (Benkard 2000), and a long-standing practical wisdom in industrial circles: do everything possible to avoid prolonged discontinuities in activity, even if it means accepting temporary overcapacity. For instance, in 2018, during a parliamentary hearing, a top executive from the French nuclear operator said: "If I were to use an image to describe our situation, it would be that of a cyclist who must keep pedaling so as not to fall." In a sense, the decisions made during recent crises, such as the economic downturn in Germany in the early 2010s (the Kurzarbeit system) or, more recently, the COVID-19 crisis, to maintain employment "whatever it takes" by resorting to furlough schemes rather than layoffs, were based on the same type of reasoning: the costs of temporary overcapacity are lower than those of rebuilding lost capacity. In any case, maintaining *ex-ante* overcapacity is a way to soften the impact of discontinuities and to make the "learning wall" less steep. Other measures can contribute as well, such as systematically facilitating overlap between employees approaching retirement and younger staff, or managing the workforce's age distribution so as to prevent overly unbalanced situations in which an abundance of inexperienced personnel overwhelms the teams' integration capacity.

3.5 Conclusion

In conclusion, one might wonder whether the French nuclear operator's experience is transferable to other industrial sectors. As we mentioned at the beginning of this chapter, the world of polycrisis (Lawrence et al. 2024) and unpredictability that seems to have become our reality may produce analogous phenomena in other industries. Perhaps these will not be as extreme as those experienced by nuclear energy: after all, we are talking about a "stop-and-go" scenario with a complete halt to new projects over a 10-year period. But other sectors could experience successive slowdowns, or even shutdowns (Tooze 2021), and restarts. One can already sense that all companies operating in a monopsonistic position, such as nuclear operators, may experience similar effects. Consider, for example, the defense industry, the manufacturing of large-scale components, or even major infrastructure projects (such as aircraft carriers or nuclear submarines), which cannot rely on a steady, guaranteed flow of projects. Yet the issue may, in fact, turn out to be even more general. This question is perhaps crucial for the future of Europe: the goal of "sovereignty," or at least of reducing certain external dependencies, will necessarily involve processes of building or rebuilding industrial capacities, whether in innovation, development, or manufacturing. For instance, what would bringing semiconductor production back to Europe imply? Reducing European dependence on Chinese steel? Very quickly, a large number of people would need to be trained in highly complex activities, starting from a relatively thin initial talent pool. And what about the widely expressed wish, in 2025, for a rapid ramp-up in the defense industry? Might we have a multitude of "learning walls" ahead of us, possibly depending on the political decisions to come?

References

L. Argote, Organizational learning curves: persistence, transfer and turnover. Int. J. Technol. Manage. **11**(7–8), 759–769 (1996)

L. Argote, D. Epple, Learning curves in manufacturing. Science **247**(4945), 920–924 (1990)

J.B. Barney, A.M. Arikan, The Resource-based View: Origins and Implications. Blackwell Handb. Strat. Manag. 123–182 (2005)

C.L. Benkard, Learning and forgetting: the dynamics of aircraft production. Am. Econ. Rev. **90**(4), 1034–1054 (2000)

B. Flyvbjerg, Public planning of mega-projects: overestimation of demand and underestimation of costs, in *Decision-Making on Mega-Projects: Cost-Benefit Analysis, Planning, and Innovation*, 120–44 (2008)

F. Garcias, C. Dalmasso, J. C. Sardas, paradoxical tensions in learning processes: exploration, exploitation and exploitative learning. M@n@gement**18**(2), 156–178 (2015)

F. Garcias, C. Dalmasso, C. Depeyre, "Can't remember what I forgot:" investigating organizational forgetting within a project-based organization. Proj. Manag. J. (online first) (2024)

F. Garcias, S. Tillement, L'industrie nucléaire civile: pour une histoire organisationnelle. Entrep. Hist. **114**(1), 6–20 (2024)

C. Gelhard, S. Von Delft, S.P. Gudergan, Heterogeneity in dynamic capability configurations: equifinality and strategic performance. J. Bus. Res. **69**(11), 5272–5279 (2016)

R.W. Grimes, W.J. Nuttall, Generating the option of a two-stage nuclear renaissance. Science **329**(5993), 799–803 (2010)

A. Grubler, The costs of the French nuclear scale-up: a case of negative learning by doing. Energy Policy **38**(9), 5174–5188 (2010)

P.M.D. Holan, N. Phillips, The remembrance of things past? the dynamics of organizational forgetting. Manage. Sci. **50**(11), 1603 (2004)

M. Lawrence, T. Homer-Dixon, S. Janzwood, J. Rockstöm, O. Renn, J.F. Donges, Global polycrisis: the causal mechanisms of crisis entanglement. Global Sustainability **7**, e6 (2024)

M. Silvi, E.P. Rosa, A tragedy of the horizons? an intertemporal perspective on public support for carbon taxes. Energy Economics **125**, 106815 (2023)

N. Slawinski, P. Bansal, Short on time: intertemporal tensions in business sustainability. Organ. Sci. **26**(2), 531–549 (2015)

A. Tooze, *Shutdown: How Covid Shook the World's Economy*. Penguin UK (2021)

K.E. Van Oorschot, H. Akkermans, K. Sengupta, L.N. Van Wassenhove, Anatomy of a decision trap in complex new product development projects. Acad. Manag. J. **56**(1), 285–307 (2013)

Chapter 4
Digital Transitions in Industrial Work

Between Skills Development and Deskilling Risks

Marta Santos

Abstract This chapter examines how digital transitions in industrial settings affect skills development, work autonomy, and collective dynamics. Drawing on two case studies from logistics operations within a multinational company in Portugal, the analysis highlights the divergent effects of technological change: in one case, technological innovation supported learning, collaboration, and the recognition of workers' experience; in the other, it led to increased control, task fragmentation, and a reduction in opportunities to mobilize existing skills. The chapter discusses the paradox between the urgent need for upskilling to face digital transitions and the simultaneous introduction of systems that limit discretionary action, weaken cooperation processes, and create risks of deskilling. Finally, the discussion considers broader implications for how transitions are managed in concrete work settings, emphasizing the need to acknowledge workers' contributions and to support the conditions under which work activity can remain a resource for development.

Keywords Digitalization · Upskilling · Deskilling · Change management

4.1 Introduction

In recent years, discussions about work have typically centered around two key topics.

The first focuses on the identification of the transformations currently underway worldwide that significantly impact how work is organized and the conditions under which it is performed. References are commonly made to climate change, rapid digitalization, political and economic tensions, geopolitical instability, shifting employment relationships, and the challenges associated with an aging population, all of which affect the workforce (e.g., Bieder et al. 2024).

M. Santos (✉)
Center for Psychology at University of Porto, Faculty of Psychology and Educational Sciences, University of Porto, Porto, Portugal
e-mail: marta@fpce.up.pt

S. Tillement et al. (eds.), *The Shaken Foundations of Safety*,
SpringerBriefs in Safety Management, https://doi.org/10.1007/978-3-032-22924-3_4

The second topic stems from the first and is presented as a way of responding to these challenges: In a rapidly changing world, workers are expected to develop the competencies required to navigate, adapt to, and engage with these transformations.

Focusing this analysis specifically on the changes arising from technological transitions, numerous official documents underscore the skills gap that threatens companies' ability to integrate new technologies into their operational frameworks. Addressing this gap appears to require actively engaging workers in upskilling and reskilling efforts, whether through initial or continuous training programs designed to update their qualifications (AAA 2017; OECD 2019; WEF 2023). The general stance is that workers must either adapt to these shifts or risk their knowledge becoming obsolete, especially as advancements in machine capabilities make some skills increasingly vulnerable.

However, are these truly new concerns?

In reality, the notion that the world is changing, and that workers' skills and knowledge must be continually updated, has been present at least since the 1990s. For instance, in the context of promoting greater social cohesion, it was noted in 2000 that the emerging knowledge-based society that Europe aspired to become carried the risk of deepening the divide between those with access to knowledge and those excluded from it. Therefore, it was deemed essential to undertake all necessary efforts to enhance the skills of European citizens (European Parliament 2000). The competencies that are highly valued today find their roots in those defined at the time. While the structure and rationale behind them may have become more refined, the emphasis on technological and digital skills, social and interpersonal abilities, and the capacity for lifelong learning has remained consistent over the decades (Commission of the European Communities 2001; WEF 2023).

One can, of course, acknowledge that the urgency now feels greater and that the pace and disruptive nature of current changes have intensified. Nonetheless, today's debates are underpinned by a longer historical trajectory and by current workplace realities that must be taken into account if we are to forecast the future of work and the competencies required to perform it accurately.

In a context where the emphasis is on updating workers' skills to meet evolving demands, it is important to understand how the skills gap is manifesting in workplaces and how upskilling and reskilling processes are being implemented. This analysis is crucial not only to identify which types of skills are becoming more valued for performing work activities and what impacts this may have on individuals and work collectives but also to assess the role that human resources teams may be playing in these processes.

To better understand these issues and ground this reflection in real work settings, this chapter analyzes two real-life cases of workplaces currently undergoing or having recently undergone technological transition.

4.2 Competence Development in Concrete Work Transition Situations

The work activities analyzed in the two cases underwent substantial transformations resulting from the introduction of AI-based systems designed to support operational processes. Both cases took place in an industrial context, specifically in the warehouses of a multinational company operating in Portugal, providing technology and services.

In Case 1, the transition was associated with the implementation of a pallet-tracking system, which required operators to engage with new digital tools and software in order to verify and label products received from suppliers.

In Case 2, the deployment of a newly digitized and optimized supply system fundamentally reconfigured the entire delivery logic: rather than adhering to a fixed, pre-scheduled route, deliveries were now dynamically determined through the integration of sensors and an algorithm that optimized delivery time, routing, and the volume and type of products transported to the production lines.

In both cases, the changes introduced were regarded by the company's internal digitalization team as highly complex processes, involving intelligent automation supported by IoT and optimization/AI algorithms, situated within the broader framework of Industry 4.0.

The analysis was conducted using qualitative and participatory methodologies grounded in the study of work activity. This approach was intended to anticipate potential forms of activity that might emerge in response to these transitions by identifying opportunities to mobilize workers' experiential knowledge and to foster skill development within work collectives.

Although both digital transitions were implemented in a similar manner, i.e., they were led by design teams in collaboration with logistics management, with limited involvement of operators and primarily restricted to late stages of testing or implementation, their impact on work activity was perceived in markedly different ways by those directly performing the tasks.

4.2.1 Transitions without Disruption: Work as a Site for Learning and Development

In Case 1, the custom-built software for verifying and labeling inbound products was perceived by logistics operators as functional and appropriate. It was simpler and more intuitive than the previous software, allowing workers to complete their tasks more efficiently and with less interaction time per task. The system did not impose a rhythm or temporal sequence for performing tasks and was not tied to performance metrics.

Importantly, operators had the opportunity to suggest improvements to the software, which were welcomed and implemented. This co-construction contributed to a sense of ownership and made learning a collective process. The work collective played a key role in facilitating peer learning, drawing on experienced workers to guide others, encouraging flexibility, and ensuring the job was done "properly," even when that meant bypassing rigid procedures.

Because the new system did not disrupt previous workflows, workers continued to use and develop their existing knowledge and skills. This included:

- cognitive competencies related to task organization and problem-solving (e.g., handling missing or unclear product information, managing space and stock priorities);
- coordination abilities to ensure the achievement of both warehouse and production goals;
- self-leadership, as workers identified improvement opportunities and learning needs;
- digital skills required to interact with the new software.

In summary, despite the technological change, workers felt they maintained the ability to work in their own way, with sufficient autonomy to respond to problems and unforeseen events.

Even in the context of substantial technological investment, this case shows that when space is given to workers to reorganize, participate in transitions, and learn from one another, it is possible to create new, high-quality, and reliable work methods.

4.2.2 Transitions that Constrain: The Risk of Deskilling Through Digitalization

In Case 2, the warehouse replaced fixed-cycle delivery routes to the production lines with a digital delivery system based on information collected through sensors and on algorithm-optimized routes that varied according to time and product demand. Workers used a PDA-based digital assistant to receive and follow real-time delivery instructions.

Once the tool was introduced, it dictated the entire workflow: workers received delivery routes via the PDA, followed step-by-step instructions (e.g., which wagons to attach, which route to follow, and at which production lines to stop), and recorded each completed step. A new route was then automatically generated, creating a continuous loop.

This meant that workers could no longer anticipate tasks; they discovered the next delivery only after completing the current one. This limited their autonomy, and many felt that the new system reduced their engagement, increased surveillance (e.g., constant location tracking), and decreased flexibility.

Importantly, operators were already familiar with digital tools such as PDAs and scanners. The issue was not the introduction of new technology itself, but the fact that it prevented planning and responsiveness. They also highlighted inefficiencies, such as:

- time wasted when two workers were unnecessarily assigned to the same area of production;
- frequent switching between types of routes (raw versus finished products), requiring equipment changes;
- system resets due to network failures, which required the re-entry of all previously completed steps.

From the workers' perspective, the activity was carried out with diminished quality, there was a sense that time was being wasted, and there was a widespread perception that digital optimization could itself be optimized.

At the social level, the system also weakened collective dynamics. Operators reported:

- a loss of informal collaboration, since the timing between routes could no longer be synchronized with warehouse colleagues responsible for preparing orders;
- an inability to respond to urgent production issues (e.g., replacing a missing item or a defective one), which had previously fostered teamwork and shared responsibility.

Although this represented a clear case of technological transition, cognitive competencies such as anticipation, problem-solving, collective contribution, and self-leadership were apparently no longer required or even usable. Work activity thus changed significantly, although neither the new organization of work nor the new tool required upskilling, not even digital upskilling, for the workers involved.

The resulting skills gap was not between workers who possessed or lacked digital skills, but rather between workers who had relevant skills and the limited scope of what the new system allowed them to mobilize.

4.3 The Paradox of Technological Change in the Workplace

The analysis of the two work situations highlights that not all technological transitions necessarily disrupt how work is done. In some contexts, workers perceive changes as relevant and valuable, enabling them to progressively incorporate new tools without invalidating their existing experience, while at the same time creating opportunities for collective learning and reorganization.

In these cases, it is through the work activity itself that people can move from potential to reality (Puey 2022). In such situations, it becomes possible to gradually appropriate the new conditions, with both individual and collective agency, to shape how work is done (Efros 2012). The activity thus becomes one with "development potential" (Di Ruzza and Lacomblez 2014).

Case 1 clearly illustrates this dynamic: workers did not lack the skills to deal with the transition. On the contrary, they already possessed a range of key capabilities, such as analytical thinking, problem-solving, and task organization; curiosity, adaptability, and lifelong learning; empathy, active listening, and teamwork; attention to quality, and the ability to adjust to unforeseen events. These competencies were concretely expressed in the work situation, developed in direct contact with the reality of the activity, requiring the mobilization of tools and representing forms of knowledge that were recognized and valued.

These skills were reinforced through participation, collaboration, and on-the-job learning within the collective, fostering more effective and reliable work methods. In this scenario, technological change contributed to development.

However, not all transitions require or foster these types of skills. There are transitions that introduce significant changes in the way work is done, forcing disruptions, demanding new ways of working, while simultaneously stripping away the meaning workers had previously attached to their activity, and undermining the recognition this work received within the collective.

Sometimes, as in Case 2, the new technology introduced led to a reconfiguration of work based on control and strict task prescription, hindering the use of skills that operators already possess. These are situations in which change is presented as desirable and inevitable, but where previous ways of working are disregarded. The experience that could otherwise be a resource is ignored, and workers quickly realize that their activity can revert to traditional problematic working frameworks (Pueyo 2022), characterized by fragmented work and control imposed by the tool. In the long term, this may lead to a "deskilling" of workers who were previously experienced in the activity.

This reinforces the idea that the choices companies make regarding the introduction of technologies do not just affect productivity goals and associated costs. The adoption of new technologies and the way they are implemented in concrete contexts ultimately determine ways of working. At the same time, they shape the (im)possibilities for workers' development.

We are thus faced with a paradox within organizations: while autonomy, proactivity, and individual responsibility are promoted as key elements for the future of work, the systems introduced may in fact constrain those very capacities. Largier (2022) poses the question: Is there not a profound contradiction in advocating responsibility and autonomy in workplaces where activity is tightly scripted, and workers are expected to rigidly follow the rules and "best practices" defined by designers? Similarly, Barcellini et al. (2024) point to the risks of reduced creativity and the standardization of thought with the widespread introduction of AI in work settings.

This paradox is even more critical in complex or high-risk sectors, where unpredictability is part of the job. In such contexts, safety depends on human capabilities such as adaptability, imagination, and improvisation (Cook 2020; SASG "Operator of the future," 2023). These skills are essential for safety management, particularly in crisis situations, as workers need to (i) understand the specific situation they are facing; (ii) recognize when the planned approach is not (entirely) suitable for the current situation; and (iii) decide to operate differently, which may involve improvisation (De la Garza 2023). Subsequently, competencies such as anticipation, situational awareness, collective intelligence, creativity, empathy, and practical wisdom are essential for safety, reliability, and effective performance in uncertain contexts.

In this sense, managing transitions, particularly technological ones, must avoid causing the deskilling of some workers. These competencies should not be reserved for a select few, and work should not be reduced to, or considered as if it were merely about, "carrying out" tasks.

Close monitoring of these transitions within companies, from the perspective of the evolution of work activity, worker development, and the related impacts on health, is therefore essential. It is not enough to claim in broad terms that workers need to be qualified, nor to design generic training programs for updating digital skills. Change management in the workplace must respect workers' experience and competencies, which have enabled them to handle both "business-as-usual" scenarios and unexpected situations.

Ultimately, the debate on future skills, careers, and workplace safety is inseparable from how organizations manage change today and the extent to which workers and their representatives are given space to contribute to and shape those transitions.

Acknowledgments This work is supported by national funds, through the Operational Competitiveness and Internationalization Programme (COMPETE 2020) [Project No. 182852; Funding Reference: SIFN-01-9999-FN-182852].

Ethics Approval This study was conducted in accordance with the principles of the declaration of Helsinki. Approval was Granted by the Ethics Committee of the Faculty of Psychology and Education Sciences of the University of Porto (29/01/2025; Ref. 2024–12-17b). Informed consent was obtained from all individual participants included in the study.

References

AAA—Association for Advancing Automation. Work in the automation age: sustainable careers today and into the future [White paper] (2017). AAA https://www.motorsystems.com/uploads/pdfs/Work-in-the-Automation-Age-White-Paper.pdf

F. Barcellini, T. Gamkrelidze, N. Greenan, A. Jolivet, M. Zouinar, Le travail et l'emploi à l'épreuve de l'IA: état des lieux et analyse critique de la littérature. Rapport de recherche financé par l'Agence d'objectifs de l'Institut de Recherches Économiques et Sociales dans le cadre d'une convention avec la CGT—Force Ouvrière (2024). https://ires.fr/wp-content/uploads/2024/07/AO-2018-3_travail_IA_2024.pdf

Bieder et al., Industrial safety in the age of "living with": Uncertainty, complexity and rising expectations, in *Cahiers de la Sécurité Industrielle (No. 2024–02)*, (Foundation for an Industrial Safety Culture, Toulouse. France, 2024). ISSN 2100-3874). https://doi.org/10.57071/420lwu

Commission of the European Communities, Communication from the commission: making a European area of lifelong learning a reality. Brussels, 21.11.2001. COM 678 final (2001). https://www.europarl.europa.eu/meetdocs/committees/cult/20020122/com(2001)678_en.pdf

R.I. Cook, How complex systems fail. Hindsight **31**, 13–16 (2020)

C. De la Garza, PEPSS PCD1, un dispositif de préparation des directeurs de crise à la prise de décision par le storytelling. In T. Meszaros (Dir.). *Crises Et Simulations: Exercices, Apprentissages,*

Modélisations, Gestion Et Anticipation Des Crises Dans L'espace Francophone. Les Éditions du CERF (2023).
R. Di Ruzza, M. Lacomblez, Ergologie, travail et développement. Quelques suggestions. Ergologia **12**, 129–145 (2014)
D. Efros, Des « approches participatives » aux processus d'empowerment: quelle autodétermination des conditions d'un développement ? Ergologia **6**, 41–114 (2012)
European Parliament, Lisbon European council, 23 and 24 march 2000, presidency conclusions (2000). https://www.europarl.europa.eu/summits/lis1_en.htm
A. Largier, Rethinking competencies in hazardous industries: Case study of the nuclear sector in France, in *Managing Future Challenges for Safety: Demographic Change, Digitalisation and Complexity in the 2030s*, SpringerBriefs in Safety Management, ed. by H. Laroche et al., (2022) https://doi.org/10.1007/978-3-031-07805-7
OECD, Preparing for the changing nature of work in the digital era, in *OECD Going Digital Policy Note*, (OECD, 2019)
V. Pueyo, Contribuer à des futurs souhaitables pour répondre aux défis de l'Anthropocène: les apports d'une Prospective du travail. Activités **19**(2), 1–38 (2022). https://doi.org/10.4000/activites.7540
SASG "Operator of the future", Industrial safety in a changing world—human, digital, new organizations: 10 key points for 2040, in *Cahiers de la sécurité industrielle (Issue 2023–04)*, (Foundation for an Industrial Safety Culture, Toulouse, France, 2023). https://doi.org/10.57071/240fut
World Economic Forum, The future of jobs report 2023: INSIGHT REPORT (2023). https://www.weforum.org/publications/the-future-of-jobs-report-2023/

Part II
Training Solutions for Skills Shortages

Chapter 5
Applying Resilience Engineering to Real-World Work

Elizabeth A. Lay

Abstract Traditional safety approaches focus on predicting and controlling risks in planned work, but in complex, high-risk environments, uncertainty, and surprise are ubiquitous. Resilience Engineering provides a framework for navigating emergent, dynamic risks. This chapter outlines three strategies to increase reliability and resilience: Embedding Continuous Operational Learning, Guided Adaptability, and Thriving in Uncertainty. The chapter details practices developed to inculcate resilient performance such as learning from Close Calls, reflective learning via After Action Reviews, and reframing risk as uncertainty using an Uncertainty Gauge. All build system-wide capacity to anticipate, monitor, respond, and learn to help people prepare for the unexpected.

Keywords Resilience engineering · Operational learning · Guided adaptability · Managing uncertainty · Sensemaking · Learning from everyday work

5.1 Introduction

Traditional safety approaches are designed to ensure known, planned work goes well. Yet *all work,* especially in high-risk environments, cannot be entirely known, planned, or proceduralized.

How do we work successfully in a volatile, uncertain, complex, and ambiguous (VUCA) world?

Working successfully in a VUCA world requires predicting and controlling critical risks where possible and being prepared to adapt in real-time to handle the emergent, dynamic risks that *will* arise. This chapter focuses on the latter: navigating emergent, dynamic risks and uncertainty that are inescapable in today's work environments and more prevalent than many acknowledge. It begins with embracing the reality that work is variable, uncertainty is always present, and surprise is a normal part of work.

E. A. Lay (✉)
Forge Works, Rochester, USA
e-mail: beth.lay@forgeworks.com

S. Tillement et al. (eds.), *The Shaken Foundations of Safety*,
SpringerBriefs in Safety Management, https://doi.org/10.1007/978-3-032-22924-3_5

"Variability of everyday performance is necessary for a system to function and is the reason for both acceptable and adverse outcomes" (Hollnagel 2014).

5.2 Concepts and Methods

Much of this chapter is based on Resilience Engineering concepts. Resilience is the ability to succeed in varying conditions. Resilience is about how well a system can gracefully extend to handle troubles that challenge boundaries of operation (Woods 2018). Resilience is the potential to act: anticipate, monitor, respond, and learn (Hollnagel 2009).

Many methods in this chapter were designed by the author and involve creating space for uncertainty to be noticed and managed. This involves intentionally approaching situations with curiosity and positioning to see as much as possible which can mean physically, as in looking from different locations, or cognitively with perspective shifts—looking from different roles, levels, or types of expertise. Creating the space for uncertainty is accomplished by asking different questions (sprinkled throughout the chapter). The methods and questions were developed for safety–critical industries and tested in power plant maintenance and line clearance tree work.

"Knowing that the world they face is complex, unstable, unknowable, and unpredictable, they position themselves to see as much as possible" (Weick and Sutcliffe 2001).

5.3 Three Key Strategies

This chapter outlines three key strategies that can increase reliability and resilience of individuals, teams, and systems: (1) embedding continuous operational learning; (2) guiding adaptability; (3) thriving in uncertainty. Specifically, it explores compensation strategies that maintain or improve performance by modifying training and fostering resilience, particularly as less-skilled people enter the workforce. These practices develop collective skills and facilitate the transfer of critical know-how.

5.3.1 Embedding Continuous Operational Learning

Learning is continuous and seamlessly embedded into daily work practices. Learning is informal, proactive, and inquiry driven. This becomes increasingly vital when high turnover is the norm. A simple but powerful mantra: "Learning is the key to keeping people safe" is at the core.

Leaders shift from telling people what to do to facilitating thought-provoking conversations and encouraging critical thinking. Their role is not to dictate but to guide—asking questions, listening, reflecting, and learning alongside their teams. Leaders embrace their own learning and foster a culture of shared learning.

People learn best through compelling stories and honest, blameless reflection on surprises and lessons, and with the support of leaders who remain deeply connected to the realities of the work.

5.3.1.1 Turning Close Calls into Learning Opportunities

Close Call reporting serves as a powerful lens into the real risks present in day-to-day work. In the author's experience, close calls are often under reported—leaving a wealth of risk-important insights underground. This chapter explores strategies to increase reporting of high-value close calls, beginning with close call mining to proactively surface followed by practices to encourage broader participation.

A Close Call can be defined as an event in which an individual was nearly injured, or property was almost damaged. These incidents occur when conditions are moving toward harm but are either interrupted by proactive intervention ("Good Catch") or mitigated by chance ("Luck Intervenes") to stop or reduce the severity of an incident.

5.3.1.2 "Mining" for Close Calls

Close Call reporting can be jump-started by asking people to share their stories of Close Calls that happened in the past. Close Call mining involves asking questions to uncover valuable lessons, sometimes by exploring extreme experiences—the worst, best, first, last, hardest, proudest, or scariest moments encountered on the job. Organizations can tailor inquiries to specific high-risk scenarios or keep questions open-ended, such as asking, "What story do you always share with a new teammate?".

Box 5.1: How to "Mine" for Close Call Stories Ask questions to surface Close Call stories such as:

– Can you share an experience where you learned a valuable lesson that helps keep you safe?

– Tell me about a time when you noticed something was "off" and you prevented something bad from happening. What tipped you off?

– Have you ever had a Close Call with danger—an experience so intense it left you shaken? What happened?

– What is your most memorable day at work—whether it was your best or worst?

– What's the most challenging task or situation you've ever faced at work?

– Can you tell me about a Close Call where *[specific event]* happened in a completely unexpected way? What was the outcome?

Pause after posing a question to give people time to reflect and recall meaningful experiences.

Encourage people to describe clues that signaled potential trouble to identify early warning signs.

Ask about sensory details—What did you see, hear, smell, or feel? —to create a vivid, memorable account when the story is shared later.

Clarify the dilemma to help listeners imagine themselves in the storyteller's shoes.

Delve deeper by asking, "What were you thinking at that moment?" or "How did you decide what to do?"

Capture the story in the person's own words.

5.3.1.3 Analyzing Close Calls for Risk Factors and Insights

Only reviewing incidents that resulted in harm misses the wealth of information provided by Close Call events. Analyzing Close Calls, particularly those with serious injury potential, is like *taking the blinders off*, revealing risks, insights, and trends that may otherwise remain hidden.

For instance, within line clearance tree trimming, one of the most frequent causes of severe injury or fatality is being struck by a falling tree or limb. A review of "struck-by" Close Calls reveals that many incidents occur during the final cut and often involve a recently added team member. This leads to two insights:

- the final cut introduces unique hazards requiring an elevated level of situational awareness;
- the addition of a new team member significantly changes team dynamics, essentially creating a "new team." Teams with a new team member need to increase "chatter," i.e., communication and coordination.

Box 5.2: How to Improve Close Calls Reporting 1. Ensure reporting is easy, simple, and efficient.

2. Ask three questions:

–What happened?

–What surprised us?

–What did we learn that could help others?

3. Use accessible technology (e.g., mobile phone with voice-to-text functionality).

4. Make learning visible. Establish a weekly discussion forum that brings operations and safety leaders together to review several recent Close Calls.

Encourage the reporters to share their story and participate in a positive, learning-focused conversation.

5. Embrace EVERYTHING that is reported as Close Calls. Avoid overly rigid definitions of what qualifies as a Close Call.

6. Recognize and appreciate those who report Close Calls. Foster a culture of trust and gratefulness by celebrating those who contribute. Reporting should not be anonymous, and there should never be punitive consequences.

Learning from Close Calls transforms an organization from reactive—responding to incidents *after* they occur—to proactive—managing risks and uncertainty before incidents occur. By surfacing and analyzing Close Call stories, organizations gain firsthand insights and a better understanding of the real risks in performing everyday work.

5.3.1.4 Learning Through Reflection: After Action Reviews

After Action Reviews (AARs) create the space for continuous learning that is essential to building expertise. Efforts often concentrate on pre-job preparations—such as pre-task briefs—yet valuable lessons from daily work are missed as teams leave for the day. This version of an AAR is intentionally quick and easy, reducing obstacles to adoption.

AARs could be conducted at the end of the workday, after new or infrequently performed tasks, when coaching new team members, or after meetings and workshops; anytime there are lessons to be learned (which is nearly always).

Box 5.3: How to Conduct an After Action Review Set the tone with an open, safe environment.

Encourage comfort. Invite the team to "grab some water, take a knee."

Ask these questions:

- What did you expect to happen? (The plan)
- What actually happened?
- ***What surprised you?***
- What went well and why?
- What can be improved and how?
- What did we learn that could help others?

Keep it concise and focused. Avoid unnecessary documentation.

Guide the discussion using open-ended questions to facilitate learning.

5.3.1.5 The Power of Asking "What Surprised You?"

Asking "What surprised you?" normalizes the practice of probing for surprises and helps people notice how often they are surprised. The answers reveal patterns that enable preparation for unexpected events. This question can be easily integrated into existing processes such as incident investigations, Close Call reporting, and AARs.

Framing the question as "What surprised you?" rather than "Were you surprised?" assumes surprises are common and triggers the brain to search for what IS surprising.

By adopting the mindset that surprise is normal, teams can be more prepared for the unexpected (Lay and Balkin 2021).

5.3.2 Guiding Adaptability

Guided adaptability enables people to make safe, informed adjustments in response to real-world conditions. Instead of rules that may not always fit, guided adaptability functions like guardrails, strategically placed in dangerous or the trickiest areas, where it's easy to take a wrong turn, offering guidance while allowing room for maneuverability.

Variability is not good or bad—it simply exists. Every work environment contains variability, and rather than resisting it, organizations must recognize its necessity. Embracing the presence and degree of variability in work is critical to successful adaptation. The ability to anticipate, adjust, and respond to changing conditions is essential. Adaptation is the foundation of safe and effective work.

Box 5.4: How to Guide Adaptation *Plan Pauses During Work.*

Pauses allow teams to reassess conditions and make necessary adjustments. These pauses could occur:

– At transition points between tasks
– Before critical steps in a process
– After a set number of repetitions for repetitive tasks
– At regular intervals for novel or high-risk work
– Ask: What is different than expected? What surprised you? What has changed?

Empower People to Adapt.

People must feel permitted and encouraged to adapt when necessary. Leaders seek areas where rules are not working as intended.

– Ask: What makes your job harder than it needs to be? Where are we operating slightly outside of policy (i.e., normalization of deviance)?

Solve Problems as a Team.

When facing a tough challenge, include outsiders to bring fresh perspectives. Challenge and cross check. Encourage critical thinking and alternative viewpoints by playing devil's advocate.

– Ask: What if ___ happens? How could this go wrong? What do you see that I might be missing?

Have Options Ready.

Prepare alternative ways to perform work more safely. This could include using different equipment, modifying methods or processes, adjusting team composition (e.g., adding more or different personnel), or postponing work until conditions improve.

5.3.3 Thriving in Uncertainty

Thriving in uncertainty begins with changing the goal from reducing to embracing uncertainty:

"To move beyond "uncertainty management," with its focus on uncertainty reduction, toward "uncertainty regulation," which includes a positive role for uncertainty creation" (Griffin 2020).

People are uncomfortable with ambiguity (Ellsberg 1961). People tend to ignore uncertainty or jump to assumptions to reduce uncertainty thus strategies include first making uncertainty visible then actively engaging with it.

Embracing uncertainty employs these strategies:

- train teams to notice uncertainty as a signal to pause, reassess, and seek more information;
- build monitoring for weak signals into systems;
- openly acknowledge and probe uncertainty by bringing it into conversations;
- have options ready to shift, flex, change course, or change mode of operating when system is nearing boundaries or weak signals arise.

Work is approached expecting change, using increasing uncertainty as a signal, then purposefully changing mode of operation by increasing vigilance, questioning, communication, and collaboration.

5.3.3.1 Reframing Risk as Uncertainty: Uncertainty Gauge

The Uncertainty Gauge supports exploring the controllability and predictability of a situation. It helps teams reframe risk by probing their ability to control a situation and questioning how uncertain they are about how a situation could play out or fail. By using Uncertainty Gauge scores, people can improve consistency in recognizing

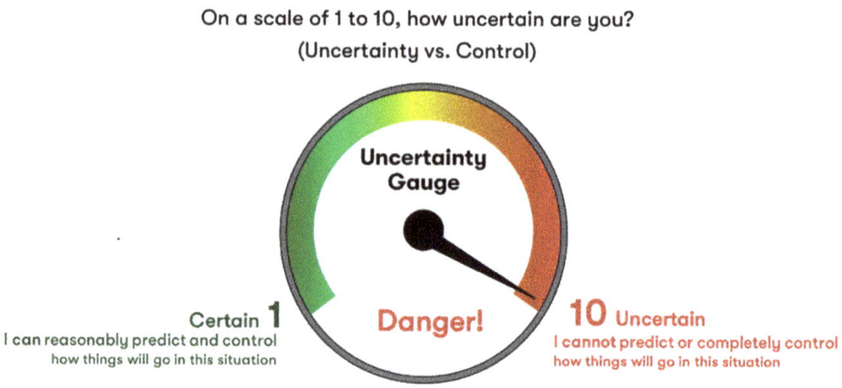

Fig. 5.1 Uncertainty gauge

when to seek a fresh perspective, when to explore options, and leaders can calibrate their team's risk tolerance (Fig. 5.1).

Todd Conklin (2017) inspired the Uncertainty Gauge creation with these statements:

- the things that kill workers are not the things with the highest perceived risk;
- the things that kill workers seem to be the things most difficult to control;
- risk is in the eye of the beholder;
- risk may be defined as the degree to which a worker is facing uncertainty.

Box 5.5: How to Use the Uncertainty Gauge Assess the Situation

– Ask: "On a scale from 1 to 10, how would you rate this situation?"

Level 1, Certain: You have no doubt about your ability to predict and control the situation, whether it's the stability of the conditions or use of tools.

Level 10, Uncertain: You have serious doubts about your ability to predict or control an aspect of the job.

Explore the Reasoning

– Ask: "Why did you give it that score?"

People will share how they would do the job, past experiences, and observations influencing their rating (e.g., what they cannot control, unfamiliar or changing conditions, or past Close Calls).

There are no wrong scores—risk perception varies by person and situation. What matters is discussing where each person lands on the gauge and why.

Uncertainty may be hard to articulate. Encourage people to "Trust your gut." If something feels off—even if they cannot explain why—raise it and explore it.

Escalate High Uncertainty

If the uncertainty score is high, escalate to next level of leadership.

Leaders establish guidelines on when they expect to be called, based on their team and nature of the work.

A leader's role is to:

– Gather input, uncover new information, and understand team concerns.

– Challenge the plan.

– Support replanning by ensuring options are available and resources are provided.

5.4 Conclusion

Resilience is a shift in perspective—an acknowledgment that surprise is inevitable, work is inherently variable, and uncertainty is ever-present. Resilience Engineering introduces practices to enable anticipating trouble ahead, managing uncertainty, and taking deliberate actions to build adaptive capacity.

The concepts, practices, tools, and—most importantly—the questions explored in this chapter help leaders foster the potential for resilience to manage emergent risks. This goes beyond traditional risk prediction and control, opening to a mindset that embraces variability, creates space for the unexpected, and embeds continuous learning into daily operations.

This chapter introduced three strategies for strengthening both reliability and resilience:

- embedding continuous operational learning;
- guided adaptability;
- thriving in uncertainty.

By integrating these strategies, organizations can build the capacity to anticipate, monitor, respond, and learn—enabling teams to adapt, expand their capabilities, and navigate uncertainty.

Acknowledgements Many practices and tools shared were invented or modified in collaboration with Asher Balkin, safety researcher, The Ohio State University Cognitive Systems Engineering Lab and with Laurin Mooney, Be Highly Reliable LLC. Concepts, tools, and practices were further developed and generalized for application across broader industries in partnership with David Provan, Forge Works.

References

T. Conklin, Workplace fatalities: failure to predict: a new safety discussion on fatality and serious event reduction. CreateSpace Indep. Publ. Platf. (2017)

D. Ellsberg, Risk, ambiguity, and the Savage axioms. Quart. J. Econ. **75**(4), 643–669 (1961). https://doi.org/10.2307/1872532

M.A. Griffin, G. Grote, When is more uncertainty better? A model of uncertainty regulation and effectiveness. Acad. Manag. Rev. **45**(4), 745–765 (2020). https://doi.org/10.5465/amr.2018.0271

E. Hollnagel, The four cornerstones of resilience engineering, in *Resilience Engineering Perspectives: Preparation and Restoration*, vol. 2, ed. by C. Nemeth, E. Hollnagel, S. Dekker (Ashgate, 2009), pp. 7–16

E. Hollnagel, *Safety-I and Safety-II: The past and future of safety management* (Ashgate Publishing, 2014)

E. Lay, A. Balkin, The second step: surprise is inevitable. Now what? in *Advancing Resilient Performance*, ed. by C.P. Nemeth, E. Hollnagel (Springer, 2021), pp. 25–36. https://doi.org/10.1007/978-3-030-74689-6

L. Mooney, One reason we fail: The "Uns." *Be Highly Reliable* (2020). Retrieved from https://www.behighlyreliable.com/read-more

D.J. Provan, D.D. Woods, S.W.A. Dekker, A.J. Rae, Safety II professionals: how resilience engineering can transform safety practice. Reliab. Eng. Syst. Saf. **195**, 106740 (2020). https://doi.org/10.1016/j.ress.2019.106740

K.E. Weick, K.M. Sutcliffe, *Managing the unexpected: Assuring high performance in an age of complexity*. (Jossey-Bass, 2001)

D.D. Woods, The theory of graceful extensibility: basic rules that govern adaptive systems. Environ. Syst. & Decis. **38**(4), 433–457 (2018). https://doi.org/10.1007/s10669-018-9709-3

Chapter 6
Tackling the "Eternal Beginners" Problem in Safety Management

The Case of Occasional Crisis Responders

Elleke Ketelaars, Martin Viktorelius, and Simon Flandin

Abstract High-hazard industries face increasing human resource challenges due to technological shifts, demographic changes, and workforce shortages, impacting crisis preparedness and safety management. Traditional vocational training often isolates skills from their work context, complicating their reintegration into professional practice. In crisis management, where unpredictability prevails, conventional training approaches focused on standardized competencies fall short. To address this, we propose two key shifts: (i) from "general skills" to "situated resilience skills," emphasizing adaptive, situated expertise over abstract competencies, and (ii) from "normal exercises" to "resilience training exercises," which prioritize improvisation and adaptiveness. Strategies such as role switching and perturbation training challenge responders to develop flexible, context-sensitive capabilities. By treating all crisis responders as "eternal beginners", training can better prepare individuals for novel threats. These changes, even if partially implemented, enhance organizational resilience and improve crisis response capacity in an era of increasing complexity and workforce instability.

Keywords Resilience · Exercises · Crisis management · Skills · Human factors

E. Ketelaars (✉) · S. Flandin
University of Geneva, Geneva, Switzerland
e-mail: elleke.ketelaars@unige.ch

S. Flandin
e-mail: simon.flandin@unige.ch

M. Viktorelius
University of Halmstad, Halmstad, Sweden
e-mail: martin.viktorelius@hh.se

S. Tillement et al. (eds.), *The Shaken Foundations of Safety*,
SpringerBriefs in Safety Management, https://doi.org/10.1007/978-3-032-22924-3_6

6.1 Introduction

In the coming years, high-hazard industries will face significant human resource challenges due to ongoing transformations such as technological advancements, digitalization, demographic shifts, and the declining appeal of industrial sectors (Laroche et al., 2022). These changes will likely result in a persistent shortage of operational staff, a skills gap, and a mismatch between available competencies and industry needs. The impact of these trends will vary across different sectors, but will have profound implications on workforce preparedness and safety management.

Consequently, the domain of vocational education and training is expected to promote approaches based on good/evidence-based practices and respond to growing demands for training in non-technical/transversal/soft skills (e.g., leadership, communication, situation awareness, stress management) and the imperative of improved performances in the short term (Flandin et al. 2022).

However, training programs often struggle to achieve satisfactory, lasting, and measurable outcomes (Bieder et al. 2017). A key challenge arises from the tendency to fragment complex human performance factors into isolated skill units extracted from their work context for organizational purposes. While this makes human factors more manageable as discrete skills, it significantly complicates the process of reintegrating these skills into cohesive and balanced professionalism.

This issue becomes particularly pronounced in crisis management scenarios, which are inherently complex, critical, and often unprecedented. It is further exacerbated when operators are required to handle such situations occasionally, rather than as part of their primary role, and in environments characterized by high turnover.

6.2 Crisis Management and Training Challenges

Crisis management is characterized by unexpectedness, uncertainty, and turbulence (Morel and Chauvin 2016; Roux-Dufort 2016; Ansell and Bartenberger 2016). The ability to react effectively and recover quickly (i.e., resilience) is essential for those involved in crisis response (Comfort and Wukich 2013; Uitdewilligen and Waller 2018). Training, particularly simulation-based training, is widely regarded as an effective method for developing crisis response capabilities (Borodzicz and Van Haperen 2002; Flandin 2022; Kleiboer 1997; Robert and Lajtha 2002). Moreover, these crisis simulations (or "exercises") are mandatory in a wide range of industrial sectors.

However, several challenges complicate crisis training:

- infrequent real-life exposure: Crisis management is not part of most responders' daily practice, making their preparedness a lower priority;
- focus on organizational testing: Exercises are often designed to assess an organization's response or to publicly demonstrate crisis readiness, rather than to foster individual, team, and organizational learning;

- inherently ill-Defined problems: Crisis situations often require collective, innovative responses based on invention and improvisation. This poses difficulties for traditional training approaches, which focus on standardized competency development at the individual level.

The combination of these challenges presents a thorny issue for vocational education and training agents in industrial sectors. They are used to teaching professionals to achieve a better mastery of well-defined problems rather than how to better cope with ill-defined ones. Such situations are already difficult to manage, but even more so in a context of personnel shortages as a result of high turnover. This raises the following question:

What are the most effective strategies for preparing crisis responders in industrial sectors for crisis management, given that the problem of turnover might result in dealing with "eternal beginners"?

This problem requires a balanced understanding of current practices, of how disadvantages can be mitigated and what alternatives exist. Based on a recent literature review on resilience training (Ketelaars et al. 2024), recent theoretical contributions in the literature, and a series of research conducted in our lab, we propose a shift in two key areas: (i) from "general skills" to "situated resilience skills," and (ii) from "normal exercises" to "resilience training exercises".

6.2.1 Shifting from "General Skills" to "Situated Resilience Skills" for Crisis Management

The emphasis on general competencies has gained traction in discussions on future workforce skills. However, the division between soft and hard skills, while heuristic, oversimplifies human factors in crisis response. Based on the aforementioned literature review on resilience training and a series of research conducted in crisis simulations and in vocational education and training, we identified five key challenges.

First, reaching a consensus: the literature on crisis management is scattered and covers a broad range of skills, including decision-making (Kapucu and Garayev 2011), communication (Johansen et al. 2012), coordination (Boin and Bynander 2015), leadership (Kapucu and Ustun 2018), and improvisation (Rankin et al. 2013; Roux-Dufort and Vidaillet 2014). This shows that there is no clear agreement on what constitutes expertise in crisis management. Second, tackling over-simplification and (re)gaining complex thinking (Poizat et al. 2025): competencies are often reduced to predefined lists, overlooking their dynamic and evolving nature. Third, considering the situated nature of practice: competencies are context-dependent; what is effective in one setting may not translate to another. Fourth, taking social interactions into consideration: crisis management relies on collaboration, which is difficult to encapsulate in individual skill assessments. Fifth, going beyond fragmented learning: extracting skills from their work context complicates reintegration into professional practice. This leads to difficulties that are hard for the learner to overcome.

Let us consider these points together and further, starting from the idea that skills are frequently isolated and extracted from their work context for organizational purposes (offers greater convenience for planning, training, and evaluation), which significantly complexifies the process of reintegrating them into cohesive and balanced professionalism for the worker. Indeed, generic competencies suffer from the fundamental problem that they are not, by conceptual implication, situated, and are thus insufficient for effective adaptation. Typically, competencies are treated as fixed skills embedded in standardized procedures, roles, and responsibilities. So, they are often seen as static, context-independent abilities that can be learned, stored, and applied across situations. Conceived as such, professional skills rely heavily on representationalist assumptions (Viktorelius 2023). By representationalism, we mean that internal models of the world that guide thought and behavior are considered essential for success. However, collective effectiveness in crisis management does not always depend on a shared representation among different agents (Pariès 2006). This challenges the notion that professional skills must be grounded in predefined cognitive models, highlighting instead the importance of adaptability and situational responsiveness.

Viktorelius (2023) proposes an enactivist ontology as an alternative theoretical basis for understanding professional skills and resilience, arguing that adaptive capacity relies upon embodied expertise and situated skillful performance. From this perspective, professional skill does not depend upon abstract, generalizable knowledge or mental models, but on situational responsiveness, expertise (phronesis), and direct engagement with context-specific contingencies. Professional skills, in the enactivist sense, are inseparable from actual practice and the lived experience of practitioners, negating the necessity of representational mediation. What is needed at the individual and team level, as proposed by Viktorelius (2023), is skillful coping and embodied know-how enacted by experienced practitioners responding to affordances in an environment to which they are constitutively coupled. Such a shift from "general skills" to "situated resilience skills" for crisis management would involve redefining adaptive capacity as inherently embodied and situated, thus opening new avenues for effective training and organizational strategies to cultivate resilient professional practices.

6.2.2 *Shifting from "Normal Exercises" to "Resilience Training Exercises"*

Classical crisis exercises are widely practiced and legally mandated in many sectors. They have proven effective in improving crisis response, enhancing resilience, and reinforcing interprofessional collaboration (Gomes et al. 2014; Hermelin et al. 2020; Pettersson et al. 2019). However, they also have notable limitations: they often serve a standardizing or evaluative function rather than fostering learning (Wybo 2008); learning outcomes and their impact on actual crisis management remain limited

(Berlin and Carlström 2015); a normalization of crises occurs (Borraz and Gisquet 2019; Wybo 2008), leading to repetitive exercises that fail to introduce new practices. Lastly, scenario writing often excludes taboo topics such as organizational dysfunctions or societal unrest, which reduces realism and challenges to teams.

Often structured and mechanistic, "normal exercises" are oriented toward familiar challenges and thus may unintentionally create vulnerabilities. Rather than preparing crisis responders for unprecedented events, traditional training reinforces confidence in procedural competence and pre-established roles. However, when faced with complex disruptions or intentionally inflicted damage, these pre-established routines can hinder rather than facilitate effective response, leading actors into "learned helplessness" (Heino et al. 2021), where reliance on familiar solutions obstructs identification of new problems and the formulation of new solutions.

To enhance their effectiveness, classical—or *normal*—exercises should be complemented with alternative training methods. We therefore propose to:

- separate exercises from training occasions, to emphasize learning over evaluation;
- diversify training formats, including:
  - video-based training;
  - low-fidelity simulations, such as table-top simulations.

Resilience training offers an alternative approach to crisis preparedness. Rooted in psychology, ecology, and material sciences, resilience is increasingly recognized as essential for socio-technical systems (Ketelaars et al. 2024). It promotes a shift in perspective: rather than viewing humans as sources of error, they are seen as valuable resources when unexpected challenges arise.

To be effective, resilience training must go beyond procedural knowledge and prepare individuals to "be unprepared", considering that "unthinkable" is not "uneducable" (Heino et al. 2021). However, resilience training is characterized by eclecticism. In the remainder of this chapter, we discuss and bring some order to the different forms of resilience training that we see as promising.

- **Resilience Training Approaches:**
  - *Psychological resilience* training aims to enhance individuals' ability to cope with challenges. Much of the existing literature focuses on resilience in the context of long-term personal hardship, such as coping with loss, trauma, or chronic stress, with the goal of improving mental health and well-being. However, this body of literature is less applicable to managing high-pressure, fast-paced situations such as crises in industrial settings. More relevant in such contexts is the literature on psychological efficacy in demanding work situations. One example is stress inoculation training. This is a form of cognitive-behavioral therapy based on the hypothesis that brief, intermittent exposure to stress enhances subsequent stress resistance (e.g., De Visser et al. 2016; Finseth et al. 2018). However, to avoid falling into the trap of seeing stress management as a generic skill, organizations should adapt training to their specific contexts to situate these skills, such as done by Andersen et al. (2016),

who grounded breathing techniques to the work of police officers in highly volatile situations.

- *Team resilience training* interventions aim to enhance the ability of a team to maintain performance when faced with known or unknown interferences. To prepare participants for unexpected situations, different types of stressors are used to put strain on teams, forcing them to reorganize their practice (see *perturbation training*). Another way to work on resilience in a team context is through role switching, also known as cross-training. The unique *Umwelt* (or subjective environment) in which professionals operate can hinder their ability to understand the needs, constraints, and responsibilities of other professions with which they must collaborate during a crisis. Role switching, or cross-training, is considered a promising method for enhancing team adaptability and flexibility (Cannon-Bowers et al. 1998; Espevik et al. 2011). Volpe et al. (1996) define it as an instructional strategy in which each team member is trained in the duties of his or her teammates. This type of training is compatible with an enactive perspective and is supported by empirical data from various other epistemological traditions.
- *Perturbation training* is regarded as a promising approach for developing resilience in high-risk organizations (Flandin et al. 2018). It involves designing situations in which the trainees' activities are deliberately disrupted, requiring them to adapt and transform their practices. Different types of perturbations can be introduced: *urgency* involves managing fast-paced episodes, leading to an acceleration of events, time pressure, heightened stress, and the risk of overwhelming decision-making and operational capacity. This type of training focuses on enhancing a team's capability to react quickly and effectively. *Indetermination* refers to varying degrees of uncertainty in situations ranging from simple unexpected events (posing a risk of surprise) to unthinkable scenarios (posing a risk of shock). This aims to develop tolerance for uncertainty and the ability to innovate under pressure. Other types of perturbation are *scarcity*, *long-duration challenges*, and *high-intensity*, each of which develops different aspects of resilience. While these principles are distinguished for conceptual clarity, they are not mutually exclusive. On the contrary, in real crisis situations they are most often combined.

6.3 Conclusion

The "eternal beginners" problem refers to the increasing challenge faced by organizations and individuals who will repeatedly experience crisis events as novices in the future, failing to effectively turn prior lessons into genuine improvements in crisis response capability. The two proposed solutions to address this problem, namely shifting organizational perspectives on crisis response competencies and rethinking preparedness training by transforming crisis exercises, can contribute to the necessary transformation of traditional preparedness training methods.

Firstly, we showed that tackling the "eternal beginners" problem requires organizations to transition from perceiving competencies as stable and universally applicable to recognizing them as emergent, context-specific, and inherently embodied, rooted in skillful adaptation rather than rigid procedural knowledge. Organizations must cultivate an awareness that crisis competency involves continuous attunement to dynamic conditions and the ability to critically question ingrained roles, effectively "unlearning" previously held certainties when confronted with novel threats.

Secondly, and more provocatively, preparedness training should change by systematically regarding all potential crisis responders in industries as first-time crisis managers, in the case of some exercises. Current exercises often reinforce existing knowledge and skill sets, conditioning actors to apply past solutions to future crises. However, severe disruptions—especially within high-risk industries—tend to produce unforeseen cascading problems, exceeding predefined capacities and revealing hidden interdependencies. Training, therefore, should be designed with scenarios explicitly constructed to challenge established protocols, forcing responders into situations with no familiar reference points. By conceptualizing every participant as a first-timer, exercises can deliberately generate uncertainty, enabling responders to critically evaluate habitual practices and dependencies on established roles, procedures, and expectations.

These two propositions align with the theoretical enactivist perspective, which argues that effective adaptability emerges from embodied coping and situated engagement, not solely from prior experience or abstract knowledge (Ketelaars et al. 2024; Viktorelius 2023). In practice, this means integrating novel training methods such as role switching, scenario building with unprecedented situations, and the implementation of pedagogical times and actions during exercises. For example, Heino et al. (2021) suggest table-top exercises in which participants design "deadlock scenarios", i.e., situations where both action and inaction lead to potentially catastrophic outcomes. This pushes trainees to confront the limitations of their typical responses and forces them to reassess their assumptions. As advocated by these authors, exposing actors to contradictions and unexpected developments tends to sensitize crisis response trainees to underlying systemic vulnerabilities and help them learn to reorganize their responses dynamically. Furthermore, designing training with the deliberate intention of "unlearning" established patterns and protocols allows actors to cultivate deeper resilience, preparing them to react flexibly and creatively in unfamiliar situations.

Of course, the organizational perspective shift toward recognizing competencies as dynamically situated (rather than standard and stable) might not be total and might not encompass the entire organization. However, if it reaches the decision makers and training designers in charge of safety development, the outcomes can be interesting. Likewise, only some of the training exercises of an organization should be designed as resilience training scenarios that go beyond the normal preparedness basis and treat responders as perpetual first-timers. The other exercises can continue to fulfill the other typical functions of assessment and reassurance. Even if implemented marginally, our propositions can be effective in promoting embodied responsiveness and strengthening crisis preparedness and organizational resilience in the face of increasingly complex and unthinkable challenges.

References

J.P. Andersen, M. Dorai, K. Papazoglou, B.B. Arnetz, Diurnal and reactivity measures of cortisol in response to intensive resilience and tactical training among special forces police. J. Occup. Environ. Med. **58**(7), e242–e248 (2016). https://doi.org/10.1097/JOM.0000000000000756

C. Ansell, M. Bartenberger, Tackling unruly public problems, in *Governance in Turbulent Times*. ed. by C. Ansell, M. Øgård, J. Trondal (Oxford University Press, 2016), pp.107–137

J.M. Berlin, E.D. Carlström, The three-level collaboration exercise–impact of learning and usefulness. J. Contingencies Crisis Manag. **23**(4), 257–265 (2015). https://doi.org/10.1111/1468-5973.12070

C. Bieder, C. Gilbert, B. Journé, H. Laroche, *Beyond Safety Training: Embedding Safety in Professional Skills* (Springer Nature, 2017)

A. Boin, F. Bynander, Explaining success and failure in crisis coordination. Geogr. Ann. Ser. B **97**(1), 123–135 (2015). https://doi.org/10.1111/geoa.12072

E.P. Borodzicz, K. Van Haperen, Individual and group learning in crisis simulations. J. Contingencies Crisis Manag. **10**(3), 139–147 (2002). https://doi.org/10.1111/1468-5973.00190

O. Borraz, E. Gisquet, L'extension du domaine de la crise ? Les exercices de gestion de crise dans la gouvernance de la filière nucléaire française. Critique internationale, N **85**(4), 43–61 (2019). https://doi.org/10.3917/crii.085.0043

J.A. Cannon-Bowers, E. Salas, E. Blickensderfer, C.A. Bowers, The impact of cross-training and workload on team functioning: a replication and extension of initial findings. Hum. Factors J. Hum. Factors Ergon. Soc. **40**(1), 92–101 (1998). https://doi.org/10.1518/001872098779480550

L.K. Comfort, C. Wukich, Developing decision-making skills for uncertain conditions: the challenge of educating effective emergency managers. J. Public Affairs Educ. **19**(1), 53–71 (2013). https://doi.org/10.1080/15236803.2013.12001720

E.J. De Visser, A. Dorfman, D. Chartrand, J. Lamon, E. Freedy, G. Weltman, Building resilience with the stress resilience training system: design validation and applications. Work **54**(2), 351–366 (2016). https://doi.org/10.3233/WOR-162295

R. Espevik, B.H. Johnsen, J. Eid, Outcomes of shared mental models of team members in cross training and high-intensity simulations. J. Cogn. Eng. Decis. Making **5**(4), 352–377 (2011). https://doi.org/10.1177/1555343411424695

T.T. Finseth, N. Keren, M.C. Dorneich, W.D. Franke, C.C. Anderson, M.C. Shelley, Evaluating the effectiveness of graduated stress exposure in virtual spaceflight hazard training. J. Cogn. Eng. Decis. Making **12**(4), 248–268 (2018). https://doi.org/10.1177/1555343418775561

S. Flandin, G. Poizat, M. Durand, Improving resilience in high-risk organizations: principles for the design of innovative training situations. Dev. Learn. Organ. Int. J. **32**(2), 9–12 (2018)

S. Flandin, Four lines of analysis for civil security crisis simulations: insights for training design, in *Simulation Training through the Lens of Experience and Activity Analysis: Healthcare, Victim Rescue and Population Protection*, ed. by S. Flandin, C. Vidal-Gomel, R. Becerril Ortega (Springer, 2022), pp. 43–61

S. Flandin, C. Vidal-Gomel, R.B. Ortega, An Introduction to simulation training through the lens of experience and activity analysis, in *Simulation Training through the Lens of Experience and Activity Analysis: Healthcare, Victim Rescue and Population Protection*. (Springer International Publishing, Cham, 2022), pp.1–23

J.O. Gomes, M.R. Borges, G.J. Huber, P.V. Carvalho, Analysis of the resilience of team performance during a nuclear emergency response exercise. Appl. Ergon. **45**(3), 780–788 (2014). https://doi.org/10.1016/j.apergo.2013.10.009

O. Heino, A. Takala, V. Huotari, Unthinkable is not un-educable: new mindset for training preparedness. Int. J. Disaster Risk Reduct. **63**, 102470 (2021)

J. Hermelin, K. Bengtsson, R. Woltjer, J. Trnka, M. Thorstensson, J. Pettersson, E. Prytz, C.-O. Jonson, Operationalising resilience for disaster medicine practitioners: capability development through training, simulation and reflection. Cogn. Technol. Work **22**(3), 667–683 (2020). https://doi.org/10.1007/s10111-019-00587-y

W. Johansen, H.K. Aggerholm, F. Frandsen, Entering new territory: a study of internal crisis management and crisis communication in organizations. Public Relat. Rev. **38**(2), 270–279 (2012). https://doi.org/10.1016/j.pubrev.2011.11.008

N. Kapucu, V. Garayev, Collaborative decision-making in emergency and disaster management. Int. J. Public Adm. **34**(6), 366–375 (2011). https://doi.org/10.1080/01900692.2011.561477

N. Kapucu, Y. Ustun, Collaborative crisis management and leadership in the public sector. Int. J. Public Adm. **41**(7), 548–561 (2018). https://doi.org/10.1080/01900692.2017.1280819

E. Ketelaars, C. Gaudin, S. Flandin, G. Poizat, Resilience training for critical situation management. an umbrella and a systematic literature review. Saf. Sci. **170**, 106311 (2024)

M. Kleiboer, Simulation methodology for crisis management support. J. Contingencies Crisis Manag. **5**(4), 198–206 (1997). https://doi.org/10.1111/1468-5973.00057

H. Laroche, C. Bieder, J. Villena-López, Managing Future Challenges for Safety: Demographic Change, Digitalisation and Complexity in the 2030s. (Springer Nature, 2022).

G. Morel, C. Chauvin, Crisis management: what are the methodological challenges for ergonomics? Le Travail Humain **79**(1), 71–96 (2016). https://doi.org/10.3917/th.791.0071

J. Pariès, Complexity, Emergence, Resilience…, in *Resilience Engineering: Concepts and Precepts*. ed. by D.D. Woods (Taylor & Francis Group, 2006), pp.43–53

G. Poizat, E. Ketelaars, J.-C. Le Coze, S. Flandin, Sécurité industrielle au prisme de la complexité: implications pour la formation. (*Industrial safety through the lens of complexity: implications for training*) Intellectica **2**(83), 59–88 (2025)

R.J. Pettersson, C.-O. Jonson, P. Berggren, R. Woltjer, J. Hermelin, E. Prytz, Resilience training of regional medical command and control. Prehosp. Disaster Med. **34**(s1), s164–s165 (2019). https://doi.org/10.1017/S1049023X19003741

A. Rankin, N. Dahlbäck, J. Lundberg, A case study of factor influencing role improvisation in crisis response teams. Cogn. Technol. Work **15**(1), 79–93 (2013). https://doi.org/10.1007/s10111-011-0186-3

B. Robert, C. Lajtha, A new approach to crisis management. J Contingencies Crisis Manag **10**(4), 181–191 (2002). https://doi.org/10.1111/1468-5973.00195

C. Roux-Dufort, B. Vidaillet, The difficulties of improvising in a crisis situation–a case study. Int. Stud. Manag. Organ. **33**(1), 86–115 (2014). https://doi.org/10.1080/00208825.2003.11043675

C. Roux-Dufort, Delving into the roots of crises: the genealogy of surprise, in *The Handbook of International Crisis Communication Research*, ed. by A. Schwarz, M. W. Seeger, C. Auer, 1st edn. (Wiley, 2016), pp. 24–33. https://doi.org/10.1002/9781118516812.ch3

S. Uitdewilligen, M.J. Waller, Information sharing and decision-making in multidisciplinary crisis management teams. J. Organ. Behav. **39**(6), 731–748 (2018). https://doi.org/10.1002/job.2301

M. Viktorelius, A critical analysis of the concept of resilience skills from an enactivist perspective, in *Human Error, Reliability, Resilience, and Performance*, ed. by R.L. Boring (AHFE International, 2023), pp. 13–20. https://doi.org/10.54941/ahfe1003549

C.E. Volpe, J.A. Cannon-Bowers, E. Salas, P.E. Spector, The impact of cross-training on team functioning: an empirical investigation. Hum. Factors J. Hum. Factors Ergon. Soc. **38**(1), 87–100 (1996). https://doi.org/10.1518/001872096778940741

J.-L. Wybo, The Role of Simulation Exercises in the Assessment of Robustness and Resilience of Private or Public Organizations, in *Resilience of Cities to Terrorist and other Threats*, ed. by H.J. Pasman, I.A. Kirillov (Springer Netherlands, 2008), pp. 491–507. https://doi.org/10.1007/978-1-4020-8489-8_23

Chapter 7
Overcoming the Paradox of Training Design for Future and Ill-Defined Practices

The Case of Nuclear Dismantling Teleoperators

Laurane De Carvalho, Raluca Ciobanu, and Germain Poizat

Abstract Nuclear decommissioning exemplifies a rapidly evolving and emerging field where new and specialized skills are required due to changing work practices. Over the next few decades, several French power plants will need to be dismantled. Most of the operations will be carried out remotely, but this has never been done before for graphite-gas reactors. A new training program for future teleoperators therefore needs to be designed. How can we train for and anticipate the skills required for a practice that does not yet exist? How can we ground training design in the analysis of work-as-done, as recommended in current guidelines, when the work itself remains hypothetical or has yet to take shape? We suggest adopting an anticipatory stance and a future-oriented approach to meet the new challenges faced by organizations. Through our ongoing empirical research in partnership with Électricité de France (EDF), we then illustrate our proposal to study existing activities using a practice-approximation approach in order to develop innovative training programs and situations.

Keywords Nuclear decommissioning · Teleoperation · Training design · Practice-approximation approach

---

L. De Carvalho (✉) · G. Poizat (✉)
University of Geneva, Geneva, Switzerland
e-mail: laurane.decarvalho@unige.ch

G. Poizat
e-mail: germain.poizat@unige.ch

R. Ciobanu
EDF Lab, Palaiseau, France
e-mail: raluca.ciobanu@edf.fr

S. Tillement et al. (eds.), *The Shaken Foundations of Safety*,
SpringerBriefs in Safety Management, https://doi.org/10.1007/978-3-032-22924-3_7

7.1 Introduction

Work is changing rapidly in today's society. In particular, we can observe a transformation of jobs as a result of digitalization and technological developments. The constant process of innovation requires the development of new means (Laroche et al. 2022). From this perspective, it is important to anticipate the "industry of the future" as well as the issues of tomorrow—including those related to safety—and seek to develop innovative training approaches to meet future skills needs.

Nuclear decommissioning exemplifies a rapidly evolving and emerging field shaped by environmental imperatives and specialized skills. Over the next few decades, several French power plants will need to be dismantled. The first planned dismantling is that of Chinon A2, a graphite-gas reactor (UNGG) built in the 1960s. Operated by EDF (Électricité de France),[1] it has been shut down since 1985. With a start date scheduled for 2030, the total duration of operations is estimated at around 30 years. According to EDF's ideal strategy, the dismantling of Chinon A2 will be followed by the dismantling of the other five UNGG reactors, starting in 2050. UNGG reactors rely on a discontinued technology, making these operations highly specialized and narrowly focused. Unsurprisingly, the dismantling of Chinon A2 will present a significant technical challenge. The dismantling of graphite-gas plants is a completely unprecedented operation, especially since some of these procedures will be carried out remotely, for safety reasons. From this perspective, developing new work methods for health and safety reasons can also introduce new risks. Furthermore, the tools for dismantling graphite reactors do not yet exist and are currently under development. They must be tested, along with all the scenarios that can be envisioned for the various stages of dismantling. Indeed, although the internal layout of the reactor is known, the condition of its structures and materials will remain uncertain until dismantling actually begins.

Teleoperation is a complex practice. The existing literature offers limited insight into the specific competencies required for teleoperation, and even less insight into how these skills can be effectively developed, particularly in training contexts (De Carvalho et al. 2026). When these skills are addressed, they are typically examined from a techno-centered perspective, overlooking the role of the human operator. It is now essential to overcome this blind spot to ensure the effectiveness and relevance of future practices and training.

Since 2022, EDF has been collaborating with the IAEA (International Atomic Energy Agency) to develop a training program for future dismantling teleoperators. The aim is to follow an activity-centered approach to identify the skills involved, from the most embodied to the most abstract, in order to develop an appropriate training program. Designing such a training program presents several challenges. The activity-centered approach is based on the assumption that training should reflect the complexity and situatedness of real work practices. A central tenet is that an empirical analysis of real work practices is a critical prerequisite to any design process, as it

[1] EDF is a state-owned French multinational electric utility company operating the 56 French nuclear power plants.

enables the development of training content and activities that are truly relevant and effective (Poizat et al. 2016). By examining activity in advance, researchers and designers can capture the embodied and lived dimensions of professional experience in real work situations to inform the design of training environments that reflect work-as-done.

In the context of UNGG dismantling teleoperation, the pivotal questions are as follows: How can we train for and anticipate the skills required for a practice that does not yet exist? How can we conduct prior analyses when the work itself has not yet been established? How can we ensure that design choices are grounded in empirical data on real-world work practices? The paradox underlying these questions is one of the central issues of our research conducted in collaboration with EDF.

The Paradox of Training Design for Future and Ill-Defined Practices

Since the recognition of the paradox of the ergonomic design process (Theureau and Pinsky 1984; Pinsky and Theureau 1987), it has been argued that the researcher or ergonomist must develop and implement methods aimed at anticipating future activity. This well-known paradox emphasizes the fact that the activity, as it will take place in the transformed situation, is inherently not fully knowable during the design process, since the effects of the intended transformations only become observable after their implementation. The earlier the contribution of the researcher/ergonomist takes place in the design process, the more effective it is, but the less empirically grounded it is. The later the contribution occurs, the more empirically grounded it is, but the less effective it becomes. Design choices play a decisive role in structuring future activity, either by opening up or foreclosing entire fields of possible action. In the context of anticipating future activity within the design process, Daniellou (2004) introduces the notion of "possible future activity," intended to anticipate the potential configuration of actual activity resulting from the choices made during the design phase. The aim is not to forecast future activity precisely, but to delineate the space of possible forms of future activity or, in other words, to evaluate how design choices may constrain or enable future actions. A key strength of this approach lies in its empirical grounding, derived from the analysis of actual work activities conducted in "reference situations," "springboards," through simulation, or through "engagement with scenarios." However, as mentioned by Biencourt (2025), anticipatory efforts are restricted to the internal scope of a "well-defined" design project and to near-term temporal spans. Despite the integration of foresight tools and an awareness of the plurality of possible futures, it cannot be said that design ergonomics fully adopts a prospective stance, defined by wide-ranging, risky, holistic, and forward-looking projections (Biencourt 2025).

Our work also echoes recent developments in prospective ergonomics (Robert & Brangier 2012, 2024) or the prospective approach to work (Pueyo 2022). Prospective ergonomics primarily focuses on products and services, on anticipating and creating for future needs and uses. As pointed out by Biencourt (2025), while the prospective dimension is embraced more explicitly, particularly through creative methodologies, the reliance on empirical data related to activity remains more ambiguous. Although the authors mention the concept of activity through various methods, they do not give it a central role. The prospective approach to work, on the other hand, focuses on desirable futures while striving to advocate for concrete utopias.

Our work adopts a middle path between design ergonomics and more prospective stances within this field (or others) to design future training programs. Our objective is not to engage in foresight in the strict sense, but rather to adopt a design-oriented approach to training situations, enriched by a prospective (or future-oriented) attitude oriented toward the plurality of possible futures of work and training activities, albeit on a distant horizon.

The remainder of the chapter explores the concept of anticipation and presents the research choices made to support the anticipation and design of training programs for "ill-defined" UNGG dismantling activities, based on empirical foundations.

7.2 An Anticipation Perspective

How can we anticipate the future skills that workers in our organizations will need, given the evolving nature of work? The evolutions may be of all kinds. In our research, they are related to the development of specific technological tools to carry out UNGG dismantling operations remotely, raising entirely new safety issues. Wilson (2013) argues that technological change is one of the main drivers of change in the pattern of demand for skills. He focuses specifically on the anticipation of future skills. While he acknowledges that the future remains unpredictable, he asserts that attempting to glimpse into it is inevitable. According to him, we can either allow individuals to form their own ideas about the future, or we must approach it in a reflective, systematic, and explicit manner, ultimately maximizing flexibility and proactivity.

Anticipation science is an invitation to consider all the different kinds of possible futures, not just the plausible, the probable, or the desirable ones (Hancock & Bezold 2017). According to Poli (2017) and Salini (2015), in considering how possible futures are intrinsically intertwined with the present, anticipation utilizes the exploration of possible futures to guide actions and inform decision-making in the present. Rather than forecasting a future as a linear extension of the past, the anticipation approach reverses the dynamic by actively generating new possibilities. The ontological hypothesis is that the past and the future coexist within the present, which becomes the realm in which they are most fully brought to life. This is why anticipation science brings the exploration of futures back to the present in order to question the necessary choices, decisions, and strategies to shape the future. The future is not merely a matter of predetermined possibilities; it is also a matter of choices and actions. The future is something that can be brought into existence. Anticipation thus has two dimensions: one oriented toward exploring futures, and the other oriented toward applying this exploration to present decision-making. Backcasting, an emblematic anticipation tool, invites us to take into account what happened in the past, envision the future, and return to the present enriched by all the considered possibilities. In our work, we more often refer to the sciences of anticipation rather than classical foresight. We do not merely imagine possible futures; rather, we seek to actively design training situations in the near present in order to foster reflection and guide the evolution of training programs in response to emerging technologies and to our understanding of teleoperation skills as it evolves throughout the research process. Moreover, from our point of view, ergonomics should be concerned more with anticipation than with mere prospective approaches.

Through our ongoing empirical research on remote dismantling operations, we aim to illustrate how we have adopted an anticipation-based, future-oriented perspective combined with a practice-approximation approach to address future skill requirements. As mentioned previously, this work is shaped by two key specificities: a particular temporal horizon—designing and studying training for an activity that is both distant in time and still largely undefined—and the challenge of addressing the needs of a profession that does not yet fully exist. The lack of established reference situations makes it especially difficult to ground the analysis in existing practices. Therefore, alternative methods are necessary to anticipate and support the development of future skills. In light of this, we have chosen to take a practice-approximation approach to ensure a minimal empirical anchorage for this anticipation and future-oriented work. The challenge is, of course, to ultimately contribute to the design of innovative training programs.

7.3 A Practice-Approximation Approach

Our grounding in the course-of-action research program and ergonomics tradition presupposes a prior activity analysis (Poizat et al., 2016) of the existing work situations to design relevant and effective training programs. The aim is to understand the activity from the actors' perspective, so as to design training programs that are attuned to the realities of their actual work. In our case, the possibilities of analyzing work-as-done are limited, if not entirely nonexistent. To date, only a single UNGG reactor[2] has been dismantled; there is no known teleoperator assigned to the Chinon A2 site; and there is no existing training for graphite-gas dismantling. This raises significant methodological questions regarding the prior specification of competencies to be targeted in training, the training programs to develop, and the learning situations to design and support. When designing training for situations where physical or time access is not possible, it is necessary to identify alternatives. In our research, our goal is to understand the future teleoperated dismantling activities by identifying the associated challenges and required skills. To accomplish this, we are going to analyze existing activities using a practice-approximation approach to anticipate the future as far in advance as possible (Duarte & Lima 2012). Grossman et al. (2009) describe practice approximations as opportunities for novices to engage in practices more or less similar to those of a profession. The purpose of practice approximations is to

[2] For which the only feedback has concerned technical aspects, with no insight into the socio-organizational or human dimension.

emphasize key aspects of practice, making it more accessible to novices. By definition, practice approximations are not the actual practice. Their characteristics differ in the following ways: (a) the *component* of the practice that is approximated, (b) the *degree of authenticity or congruence* with which the activity approximates the actual practice, and (c) the *level of complexity* with which the practice is approximated. Finding a good approximation of practice is a decisive factor. This is true both for novices training to enter their future profession and for research aimed at producing "springboard data" for design. According to Haué (2003), springboard data refers to observations of moments relevant to the envisaged solution. Such observations may involve the evaluation of existing technologies or situations in the field of application. However, they may also concern situations or technologies from different fields, providing they are comparable to those envisaged, particularly to explore the various possible variations of the studied object. Springboard data can be used to consider the future activity or to assess the relevance of hypotheses.

In the context of our research on teleoperator skills for nuclear dismantling, we have decided to conduct empirical observations in real-world settings to anticipate the upcoming activity of the future teleoperators as accurately as possible. Based on our three ongoing empirical studies, we will show how analyzing activity in proxies can generate "springboard data" to anticipate future activity, identify the related skills, and design training programs for an activity that does not (yet) exist.

7.4 Finding Good Proxies to Understand an Inaccessible Activity

Our research is based on three distinct empirical studies in different contexts. All the situations studied are considered proxies of future teleoperation situations in the dismantling process. These studies aim to provide insights into the real work of teleoperators, help us identify what is required for dismantling, and guide the development of training situations and programs.

7.4.1 Proxy 1: Test Technicians Developing Teleoperation Skills

To find a way to understand the nature of future dismantling teleoperation activity and the issues involved in the various stages of dismantling graphite-gas reactors, we considered it necessary to observe a situation approximating the dismantling of UNGG reactors as closely as possible. The *real-work situations of test technicians* using teleoperation to test dismantling scenarios for the Chinon A2 reactor are the closest to those that will be experienced by the future dismantling teleoperators. These technicians work in a facility entirely dedicated to developing and testing potential dismantling tools, methods, and processes. All the operations centralized and carried out there are designed to assess the feasibility of the dismantling scenarios under normal and degraded conditions. The technicians sometimes use teleoperation to test these scenarios, to get as close as possible to the conditions that future dismantling operators will experience. Thus, we hypothesize that it is possible to anticipate skill requirements by studying the work of the technicians who currently use teleoperation to test Chinon A2 dismantling scenarios. Although the tools to be used are still being designed, the technicians are currently using very similar ones. Analyzing their activity in this initial empirical study can therefore help document the necessary skills and inform the training of future teleoperators.

7.4.2 Proxy 2: Pilot Training for Robotic Operations in Nuclear Accident Situations

No training in teleoperation for the dismantling of graphite-gas reactors has been conducted to date. However, since a training program must be designed for future teleoperators dismantling UNGG reactors, it is necessary to examine existing training situations for teleoperation activities, to determine the requirements of such programs. The proxy examined in our second empirical study focuses on *real-world teleoperator training situations in the context of robotic interventions in nuclear accidents*. This study is being conducted within a French nuclear emergency response group. Since this group specializes in interventions during major radiological accidents, the pilots operate and are trained on robots designed for diagnostic and/or intervention purposes. The pilots must regularly update their knowledge and skills through ongoing training. Although these robots are obviously different from those envisioned for dismantling, we hypothesize that documenting real training situations for teleoperation activities can provide valuable insights into the activities (and learning) of trainers and trainees in such contexts, as well as the associated challenges. Through this study, we aim to understand the knowledge mobilized by the trainer, his concerns, and what he considers significant in this context. By analyzing the trainees' training experience, we aim to understand the knowledge they acquire, the difficulties they encounter, and how the training situations expand (or do not expand) the way they teleoperate. Consequently, analyzing existing training situations can reveal valuable learning modalities and contribute to the documentation of training for future teleoperators tasked with dismantling UNGG reactors.

7.4.3 Proxy 3: Experimentation with a Training Situation in Teleoperation and Robotics

Based on the results of the previous studies, the final empirical study will focus on designing a training experiment *for future teleoperators engaged in teleoperation and robotics activities for dismantling*. The experiment is the only way to get as close as possible to a situation similar to the one the future teleoperators may encounter during training in teleoperation activities specifically for UNGG dismantling. The aim of this final study and approximation of practice is to anticipate a potential training course for future teleoperators. The experiment will be a way to test certain design principles for a training pathway dedicated to teleoperation and robotics activities for (future) teleoperators. We will analyze the participants' experience, their concerns, the difficulties they encounter, and the skills they acquire during the training. In this final study, we also seek to test the hypothesis that analog situations may serve as effective training tools for dismantling teleoperators. This hypothesis is based on the added value of using analogs as alternatives to train people for situations to which we do not have physical or temporal access. As an example, astronauts use such methods for training, as they need to prepare for space missions without actually being in space (Suedfeld 2010). To some extent, cave environments are an exceptional analog of space (Sauro et al. 2021). The CAVES (Cooperative Adventure for Valuing and Exercising human behavior and performance Skills) training program developed by the European Space Agency[3] uses cave exploration to simulate spaceflight conditions, focusing on safety, risk management, teamwork, and life in isolation. It recreates many of the demanding conditions and specific characteristics encountered during long-duration spaceflight, such as an unfamiliar environment on Earth, lack of time references, sensory deprivation, and limited communication with the outside world.

These approximations of practice will make it possible to empirically document training for the new role of dismantling teleoperators—which does not yet exist—and, through this effort of approximation, contribute to the development of a teleoperation training program (Fig. 7.1).

7.5 Conclusion

Nuclear decommissioning illustrates a rapidly evolving and emerging field where the evolutions call for new and specialized skills. Our research shows that future dismantling teleoperators will need extensive training to perform this novel practice. Anticipating future skills remains a significant, complex, and daring challenge. This is true not only for nuclear dismantling, but for every occupation and organization. In this chapter, we presented different possible pathways to help keep pace with societal and organizational transformations and to identify the skills needed for the

[3] https://www.esa.int/Science_Exploration/Human_and_Robotic_Exploration/CAVES_and_Pangaea

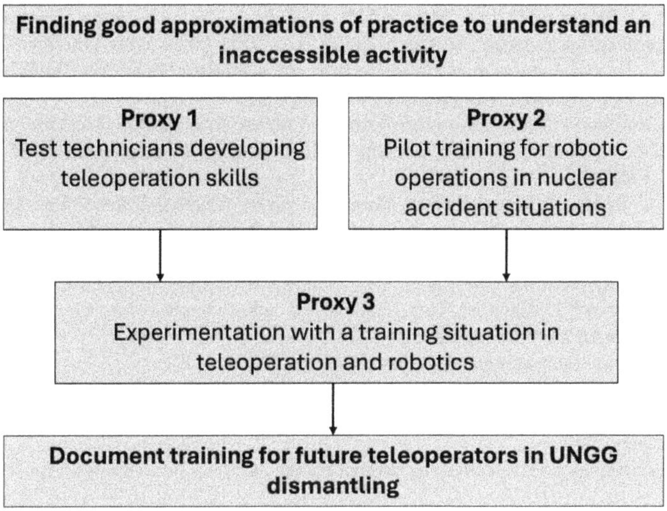

Fig. 7.1 Links between the three empirical studies and proxies

future. We suggested adopting an anticipation-type or future-oriented approach in order to envisage a whole range of possible futures. With such an approach, it is possible to make decisions in the present with creativity and proactivity, rather than merely enduring continuity with the past. Moreover, one way to anticipate the skills of the future for a given occupation is to adopt a practice-approximation approach and generate "springboard data," especially in cases where the activities do not (yet) exist. Empirical activity analysis provides added value by thoroughly documenting the skills currently mobilized in such situations of approximation or an analog, thereby taking a step toward understanding the reality of the future work. Activity analysis and anticipation-type approaches are useful for research but can also be valuable for companies and HR departments seeking to anticipate the skills of the future. These are avenues to be considered in the light of the upcoming changes in our organizations.

References

M. Biencourt, L'analyse du travail et de l'activité: socle d'une ergonomie prospective face aux enjeux de l'IA—Le cas des acteurs des centres de tri des déchets. Thèse de doctorat en Ergonomie—Université de Toulouse (2025)

F. Daniellou, L'ergonomie dans la conduite de projets de conception de systèmes de travail, in *Ergonomie*, ed. by P. Falzon (Presses Universitaires de France, 2004), pp. 359–373

L. De Carvalho, C. Gaudin, F. Palaci, R. Ciobanu, G. Poizat, *Teleoperators' Training and Skills Development: A Scoping Review* (Le Travail Humain, 2026)

F. Duarte, F. Lima, Anticiper l'activité par les configurations d'usage: proposition méthodologique pour conduite de projet. Activités **9**(9–2) (2012)

P. Grossman, C. Compton, D. Igra, M. Ronfeldt, E. Shahan, P.W. Williamson, Teaching practice: A cross-professional perspective. Teach. Coll. Rec. **111**(**9**), 2055–2100 (2009)

T. Hancock, C. Bezold, Futures thinking and healthy cities. Healthy Cities: The Theory, Policy, and Practice of Value-Based Urban Planning, 449–462 (2017)

J.B. Haué, *Conception d'interfaces grand public en terme de situations d'utilisation: le cas du Multi-Accès*. Thèse de doctorat. Université de Technologie de Compiègne (2003). https://doi.org/10.13140/RG.2.2.19183.61601/1

H. Laroche, C. Bieder, J. Villena-López, *Managing Future Challenges for Safety: Demographic Change, Digitalisation and Complexity in the 2030s* (Springer Nature, 2022)

L. Pinsky, J. Theureau, *L'étude du cours d'action: analyse du travail et conception ergonomique*. Rapport du Laboratoire de Physiologie du travail et d'Ergonomie 88, Paris: CNAM (1987)

G. Poizat, M. Durand, J. Theureau, The challenges of activity analysis for training objectives. Le Trav. Hum. **79**(3), 233–258 (2016)

R. Poli, *Introduction to Anticipation Studies* (Springer, Cham, 2017)

V. Pueyo, Contribuer à des futurs souhaitables pour répondre aux défis de l'Anthropocène: les apports d'une Prospective du travail. Activités. 19–2, Article 19–2 (2022). https://doi.org/10.4000/activites.7540

J.M. Robert, E. Brangier, Prospective ergonomics for the design of future things. Ergonomics, 1–18 (2024)

J.-M. Robert, É. Brangier, Prospective ergonomics: Origin, goal, and prospects. Work **41**(Suppl 1), 5235–5242 (2012). https://doi.org/10.3233/WOR-2012-0012-5235

D. Salini, Le conseil en évolution professionnelle et les dynamiques d'anticipation, in *Le Conseil en Evolution Professionnelle. L'activité des bénéficiaires et le métier des conseillers: deux ans d'expérience en Bourgogne*, ed. by P. Mayen (dir) (Dijon: Raison et Passions, 2015), pp.242–268

F. Sauro, J. De Waele, S. J. Payler, M. Vattano, F. M. Sauro, L.Turchi, L. Bessone, Speleology as an analogue to space exploration: The ESA CAVES training programme. Acta Astronaut. **184**, 150–166 (2021)

P. Suedfeld, Historical space psychology: Early terrestrial explorations as Mars analogues. Planet. Space Sci. **58**(4), 639–645 (2010)

J. Theureau, P. Pinsky, Paradoxe de l'ergonomie de conception et logiciel informatique. Rev. Cond. Trav. **9**, 25–31 (1984)

R. Wilson, Skills anticipation—The future of work and education. Int. J. Educ. Res. **61**, 101–110 (2013)

Part III
Managerial Solutions for Skills Shortages

Chapter 8
Skills and Careers by 2040

Emerging Challenges and Skills-Based Solutions to Foster Employability of Workers

Annet de Lange

Abstract The landscape of professional skills and industrial safety is undergoing profound changes driven by technological advancements, demographic shifts, and evolving workforce expectations. These transformations are creating significant challenges for industries, particularly those in high-hazard sectors, which must adapt their human resource (HR) strategies and safety practices to remain competitive and safe. By 2040, traditional models of skills acquisition and career development are expected to be insufficient. This chapter aims to provide a relevant overview of earlier research on the intersection of skills and career management, while proposing skills-based solutions to mitigate the challenges emerging in the global labor market.

Keywords Skills ecosystems · Sustainable employability · Hard- and soft-skill mismatch · Lifelong learning and reskilling · Aging workforce dynamics

8.1 Overview of Earlier Research

Recent research on skills ecosystems and the employability of workers highlights the critical interplay between skills development, labor market dynamics, and the evolving demands of the economy.

A. de Lange (✉)
Berenschot, Utrecht, The Netherlands
e-mail: a.delange@berenschot.nl

Open University of the Netherlands, Heerlen, The Netherlands

Norwegian University of Science and Technology, Trondheim, Norway

University of Stavanger, Stavanger, Norway

S. Tillement et al. (eds.), *The Shaken Foundations of Safety*,
SpringerBriefs in Safety Management, https://doi.org/10.1007/978-3-032-22924-3_8

8.1.1 Skills and Employability

I will start by discussing insights from earlier research on skills, employability, and aging.

A comprehensive definition of "skills" encompasses the range of abilities that individuals acquire through experience and education, enabling them to perform tasks effectively in various contexts. According to the World Health Organization (WHO), life skills are defined as "abilities for adaptive and positive behavior, that enable individuals to deal effectively with the demands and challenges of everyday life" (Darlington-Bernard 2023). This definition highlights the adaptive nature of skills, emphasizing their role in navigating real-world challenges. In the context of education, skills are often linked to academic success and personal development. The concept of skills extends into the workforce, where employers seek a combination of technical and behavioral attributes.

The concept of sustainable employability encompasses a range of skills and competencies that enhance an individual's ability to secure and maintain employment across time (De Lange et al. 2022). Alcover and colleagues argue that sustainable employability refers to "dynamic elements that influence an individual's ability to sustainably maintain employability, health, and well-being throughout the working life" (Alcover et al. 2021, p. 158).

Employability skills can be broadly categorized into two types: hard skills, which include technical and specialized knowledge, and soft skills, which encompass interpersonal abilities and personal attributes. Research indicates that employers prioritize a set of core employability skills that are essential for entry-level positions. Skills shortages in various sectors often reflect a mismatch between the skills possessed by the workforce and those demanded by employers, including both hard and soft skills (Green et al. 1998).

On the hard skills side, numeracy and technical skills are critical components of employability. Durrani and Tariq discuss how numeracy skills are increasingly being assessed during recruitment processes, indicating that a lack of these skills can significantly hinder a graduate's employability (Durrani and Tariq 2012). The work of Okoye and Nkanu further supports this, emphasizing the need for technical skills among Technical and Vocational Education and Training (TVET) graduates to enhance their effectiveness in industrial roles (Okoye and Nkanu 2020).

On the soft skills side, Song et al. emphasize the importance of personal integrity and dependability, highlighting that these traits often outweigh specific job-related skills in employer evaluations (Song et al. 2011). Rasul and Mansor echo this sentiment, noting that employers are increasingly looking for proactive employees who can take initiative and contribute to business development (Rasul and Mansor 2013). Moreover, the importance of soft skills has been underscored in various studies. For example, Thi et al. found that communication, problem-solving, and decision-making skills are among the most sought-after by employers, particularly in the context of English as a Foreign Language education (Thi et al. 2022). This aligns with the findings of the American Management Association, which states that critical thinking, creativity, and collaboration are becoming essential in today's global economy (Suarta et al. 2017). The integration of these skills into educational curricula is crucial, as highlighted by Abdullah et al., who argue for a constructive alignment between employability skills and learning outcomes in training programs (Abdullah et al. 2022).

A significant finding is that the rapid advancement of technology, particularly automation and digitalization, necessitates a substantial reskilling of the workforce. The McKinsey Global Institute projects that, by 2030, approximately 375 million individuals globally will need to acquire new skills as their current jobs evolve or become obsolete due to automation (McKinsey 2017, 2024). This shift underscores the importance of integrating STEM (Science, Technology, Engineering, and Mathematics) skills with creativity, critical thinking, and problem-solving abilities to meet the demands of the new labor market (McKinsey 2017, 2024).

The evolving landscape of the job market also necessitates that higher education institutions adapt their curricula to better prepare students and aging workers for sustainable employment. Jackson points out that employers increasingly expect graduates to possess a blend of both academic qualifications and practical employability skills (Jackson 2013). Ngulube echoes this, stressing the importance of aligning educational outcomes with labor market demands to enhance graduate employability (Ngulube 2020).

8.1.2 Integrating Perspectives on Sustainable Employability: Time and Aging

Recent research emphasizes the importance of considering time as a critical factor in sustainable employability that has been largely overlooked in earlier studies. De Lange et al. (2022) argue that sustainable employability should be viewed through a temporal lens, recognizing that both individual capabilities and job demands evolve over time. This perspective is essential for developing strategies that not only address current skills gaps but also anticipate future needs as industries and technologies continue to evolve.

The integration of temporal perspectives into employability research highlights the dynamic nature of career success and the need for ongoing skills development throughout a worker's life. For instance, the concept of "temporal leadership" has been identified as a crucial factor in managing evolving work demands in a way that supports long-term employability. Temporal leadership refers to a leader's time management behaviors that assist employees in utilizing their time at work effectively. It involves task-oriented actions like scheduling, synchronizing, and allocating time-related resources to address challenges such as temporal ambiguity, conflicting time demands, and time scarcity. Similarly, organizational time norms (e.g., expected workday and time schedules) play a significant role in shaping how employees balance work and personal life, which in turn affects their long-term career sustainability.

Moreover, the aging workforce further complicates these time-related challenges. Older employees often struggle to adapt to new technologies, and it is becoming increasingly difficult to transfer knowledge to younger generations (De Lange et al. 2021). Additionally, the career aspirations of younger workers, who value flexibility and meaningful work, create a generational divide in expectations and work styles, exacerbating skills shortages in critical areas (OECD 2023). These dynamics highlight the need for innovative HR strategies and skills development programs to maintain operational efficiency and safety, since the latter is deeply embedded in the skills and competencies of workers (Bieder et al. 2018, 2024; Laroche et al. 2021).

8.2 Skills-Based Solutions for Future Challenges

Given the anticipated skill shortages and increasing complexity of industrial operations, there is a growing consensus among experts that a skills-based approach is essential for the future.

8.2.1 Skills Passports and Lifelong Learning

One innovative solution is the development of "skills passports," i.e., digital records documenting an individual's competencies, certifications, and experiences across various roles and industries. This tool not only enhances workers' employability but also ensures their skills remain relevant in a rapidly changing job market.

The concept of skills passports is closely aligned with the need for lifelong learning and continuous professional development. By formalizing and tracking skills development through digital platforms, organizations can better manage employee career progression and ensure they are equipped to meet future demands.

HR departments will play a critical role in forecasting skills shortages and implementing strategies to address them. Research by Pak et al. (2023) emphasizes that opportunity-enhancing HR practices, such as job crafting and psychological empowerment, are particularly effective in improving the work ability and motivation of older workers. These practices, combined with the systematic use of skills passports, could help address the challenges posed by an aging workforce and ensure the retention of critical skills within organizations.

Moreover, fostering collective skills and encouraging knowledge transfer through innovative training systems are essential. Duchek (2020) stresses the importance of restructuring work to promote teamwork and collaboration, which can enhance organizational resilience and adaptability in the face of skill shortages.

8.2.2 *Creating Skills-Enhancing Ecosystems: A Multi-level, Multi-stakeholder Perspective*

In order to effectively address the complex challenges of skills development, time-related changes and career sustainability, it is crucial to adopt a more complex ecosystemic perspective that integrates multiple levels and stakeholders in the implementation of skills-based programs across time. Indeed, the development of skills is not isolated within individual organizations. It is part of a broader system that includes educational institutions, industry partners, government bodies, and workers themselves. It requires the collaboration of these various stakeholders and the alignment of their needs and goals.

8.2.2.1 Ecosystem-Based Approaches to Skills Development

The concept of skills ecosystems, defined as regional or sectoral social formations where human capabilities are developed and utilized for productive purposes, is crucial in understanding how skills are cultivated and deployed (Giacometti and Cuadrado 2020).

These ecosystems are influenced by various factors, including institutional frameworks, educational systems, and collaboration among stakeholders such as employers, training providers, and government entities (Shrestha 2024). For instance, effective labor market information systems that engage multiple stakeholders are essential for aligning training programs with market needs, thereby enhancing the employability of workers (Shrestha 2024). Moreover, the role of educational institutions in fostering local skills development is emphasized, as they are integral to the broader community ecosystem and contribute to achieving sustainable development goals (Jolliffe et al. 2023).

Furthermore, the effectiveness of skills-based, multi-level interventions is enhanced when these are designed to address the complex interdependencies between different levels of the system—individual, team, organizational, and societal. A multi-level approach allows for the consideration of how interventions at one level may impact or be impacted by conditions at another level (De Lange et al. 2024). For instance, a training program aimed at enhancing individual skills may need to be supported by organizational policies that promote knowledge sharing and collaboration across teams.

8.2.2.2 The Role of Stakeholders in Ecosystem-Based Skills Development

The involvement of multiple stakeholders in designing and implementing skills-enhancing interventions is in other words a critical basis for success. Forbes (2024) highlights the importance of managing talent ecosystems through a holistic, skills-based vision that integrates the efforts of HR departments, educational institutions, industry partners, and government agencies. Such a comprehensive approach ensures that skills development is aligned with the evolving needs of the labor market and that workers are supported throughout their careers.

For example, by involving educational institutions in the design of training programs, organizations can ensure that the skills being taught are relevant to the needs of the industry. Similarly, collaboration with government agencies can help to align workforce development initiatives with broader economic policies, ensuring that interventions are sustainable and scalable.

Furthermore, the success of ecosystem-based approaches depends on the active engagement of workers themselves. As De Lange et al. (2024) emphasize, interventions are more likely to be effective when workers have a sense of ownership and are actively involved in the design and implementation process. This can be achieved through participatory approaches that encourage workers to contribute their insights and feedback, ensuring that interventions are responsive to their needs and concerns.

8.2.2.3 Challenges and Opportunities in Ecosystem-Based Skills Development

While ecosystem-based approaches to skills development offer significant potentialities, they also present challenges. One of the key challenges is ensuring coordination among the various stakeholders involved. This requires effective communication and collaboration mechanisms that allow for the alignment of goals and the sharing of resources. Additionally, there may be conflicting interests among stakeholders, which can complicate the design and implementation of interventions.

However, these challenges also present opportunities for innovation. By leveraging the diverse perspectives and expertise of different stakeholders, organizations can develop more comprehensive and effective interventions. Moreover, the use of digital platforms and tools can facilitate coordination and communication, making it easier to manage complex, multi-stakeholder initiatives.

8.3 Avenues for Future Research and Implementation

The future of skills-based perspectives on careers and employability offers fertile ground for continued research. Below, I briefly highlight several key avenues for future investigation, drawing on insights from recent literature.

Future research should particularly explore:

- the interplay between individual agency, organizational support, and broader labor market trends in shaping **sustainable careers in a dynamic labor market**, across different life stages and career phases (De Vos and Van der Heijden 2017);

- how to **integrate employability and inclusivity**, namely how employability can be enhanced for all workers, including those in precarious or non-standard forms of employment (Hughes et al. 2021);
- the **role of organizational and institutional support** in fostering sustainable careers, and how organizations can design HR policies that not only enhance individual career outcomes but also contribute to the sustainability of the labor market as a whole, as work in practice reveals that alignment between HR practices and broader societal goals is essential for promoting sustainable employability;
- the effectiveness of various **lifelong learning and skills development initiatives**, including formal education, on-the-job training, and informal learning, in supporting sustainable careers, and how these initiatives can be tailored to meet the unique needs of different demographic groups, including older workers, young entrants, and those in mid-career transitions;
- the effectiveness of **multi-level, multi-stakeholder interventions in creating skills-enhancing ecosystems**; the identification of best practices for stakeholder involvement, communication, and collaboration to ensure that interventions are aligned with the broader goals of the ecosystem; and the role of digital platforms in facilitating these complex interventions (De Lange et al. 2024);
- the identification of **systemic barriers to skills development** and career sustainability, and the development of strategies to overcome them, ensuring that skills-enhancing interventions are effective and sustainable in the long term.

8.4 Conclusion

The future of skills and careers by 2040 will be shaped by the complex and unpredictable interplay of technological advancements, demographic changes, and evolving workforce expectations. High-hazard industries, in particular, will face significant challenges in maintaining a skilled and adaptable workforce. By adopting ecosystem-based approaches that integrate multi-level, multi-stakeholder perspectives, organizations can create environments that support continuous skills development and career sustainability as well as help individuals and organizations to respond resiliently to unpredictable developments on the labor market. Continued research and innovation in these areas will be essential to navigating the complex landscape of skills and careers in the coming decades.

References

W. Abdullah, K. Salleh, N. Sulaiman, M. Kamarrudin, Employability skills in the TVET trainer training program: the perception between experienced trainers and novices trainers. J. Techn. Educ. Train. **14**(1) (2022). https://doi.org/10.30880/jtet.2022.14.01.013

C.-M. Alcover, G. Mazzetti, M. Vignoli, Sustainable employability in the mid and late career: an integrative review. Rev. Psicol. Trab. Organ. **37**, 157–174 (2021)

C. Bieder, C.B. Benoît Journé, H. Laroche, *Beyond Safety Training-Embedding Safety in Professional Skills*. FonCSI. Springer Open (2018). https://doi.org/10.1007/978-3-319-65527-7

C. Bieder, R. Amalberti, J. Pariès, H. Laroche, E. Marsden, C. Kamaté, *Industrial Safety in the Age of "Living With"*. FonCSI (2024). https://www.foncsi.org/sites/default/files/Publications/Publications_EN/CSI_2024-02_safety-living-with.pdf

A. Darlington-Bernard, Defining life skills in health promotion at school: a scoping review. Front. Public Health **11** (2023). https://doi.org/10.3389/fpubh.2023.1296609

A.H. De Lange, K. Teoh, B. Fleuren, M. Christensen, A. Medisauskaite, L.T. Løvseth, L. Solms, A. Reig-Botella, E. Brulin, S.T. Innstrand, R. Lundmark, I. Correia, Opportunities and challenges in designing and evaluating complex multilevel, multi-stakeholder occupational health interventions in practice. Work Stress **38**(4), 352–372 (2024). https://doi.org/10.1080/02678373.2024.2332169

A.H. De Lange, B. Van der Heijden, T. Van Vuuren, T. Furunes, C. De Lange, J. Dikkers, Employable as We Age? A systematic review of relationships between age conceptualizations and employability. Front. Psychol. **12**, 605684 (2021). https://doi.org/10.3389/fpsyg.2020.605684

A.H. De Lange, D.T.A.M. Kooij, T. Furunes, What about the factor time in sustainable employability research? An overview of theory-based organizational research. Sustainability **14**(17), 10730 (2022). https://doi.org/10.3390/su141710730

A. De Vos, B.I.J.M. Van der Heijden, Current thinking on contemporary careers: the key roles of sustainable HRM and sustainability of careers. Curr. Opin. Environ. Sustain. **28**, 41–50 (2017). https://doi.org/10.1016/j.cosust.2017.07.003

S. Duchek, Organizational resilience: a capability-based conceptualization. Bus. Res. **13**(1), 215–246 (2020)

N. Durrani, V. Tariq, The role of numeracy skills in graduate employability. Educ. Train. **54**(5), 419–434 (2012). https://doi.org/10.1108/00400911211244704

Forbes Human Resources Council, *Managing Your Talent Ecosystem: A Skills-Based Holistic Vision*. Forbes (2024). https://www.forbes.com/councils/forbeshumanresourcescouncil/2024/02/20/managing-your-talent-ecosystem-a-skills-based-holistic-vision/

A. Giacometti, A. Cuadrado, *Skills Supply and Governance in the Bothnian Arc Cross-Border Region* (2020). https://doi.org/10.6027/r2020:10.1403-2503

F. Green, S. Machin, D. Wilkinson, The meaning and determinants of skills shortages. Oxford Bull. Econ. Stat. **60**(2), 165–187 (1998). https://doi.org/10.1111/1468-0084.00093

D. Hughes, C. Warhurst, E. Benger, M. Ifans, Building better futures: decent work, inclusion and careers support services in the UK. Br. J. Guid. Couns. **49**(2), 213–227 (2021). https://doi.org/10.1080/03069885.2021.1900537

D. Jackson, Student perceptions of the importance of employability skill provision in business undergraduate programs. J. Educ. Bus. **88**(5), 271–279 (2013). https://doi.org/10.1080/08832323.2012.697928

P. Jolliffe, H. Collins, J. Gold, Skilling marginalised or skills eco-system, precarious workers and higher education—A case study of Roma in Liverpool, UK. Int. J. Train. Dev. **28**(1), 22–44 (2023). https://doi.org/10.1111/ijtd.12303

H. Laroche, C. Bieder, J. Villena-López, *Managing Future Challenges for Safety Demographic Change, Digitalisation and Complexity in the 2030s* (Springer, 2021). https://link.springer.com/book/. https://doi.org/10.1007/978-3-031-07805-7

McKinsey & Company (2017). https://www.mckinsey.com/featured-insights/future-of-work/jobs-lost-jobs-gained-what-the-future-of-work-will-mean-for-jobs-skills-and-wages

McKinsey & Company (2024). https://www.mckinsey.com/featured-insights/sustainable-inclusive-growth/charts/job-skills-of-the-future

B. Ngulube, Undergraduate economics curriculum and employability skills in South Africa. Probl. Educ. 21st Century **78**(6), 1000–1013 (2020). https://doi.org/10.33225/pec/20.78.1000

OECD (2023). https://www.oecd.org/en/topics/skills-strategies.html

K. Okoye, S. Nkanu, Employers' identification of skills needed by technical and vocational education graduates for industrial work effectiveness. J. Educ. Soc. Behav. Sci. 32–41 (2020). https://doi.org/10.9734/jesbs/2020/v33i230200

K. Pak, T.A.M. Kooij, A.H. De Lange, S. Van den Heuvel, M.J.P.M. Van Veldhoven, Successful ageing at work: the role of job characteristics in growth trajectories of work ability and motivation to work amongst older workers. Acta Physiol. (Oxf) **239**, 104012 (2023). https://doi.org/10.1016/j.actpsy.2023.104012

M. Rasul, A. Mansor, Employability skills indicator as perceived by manufacturing employers. Asian Soc. Sci. **9**(8) (2013). https://doi.org/10.5539/ass.v9n8p42

R. Shrestha, Labor market information system: a review of some practices. J. Techn. Vocat. Educ. Train. **18**(1), 10–19 (2024). https://doi.org/10.3126/tvet.v18i1.62739

E. Song, D. Zhang, J. Pacha, Employability skills valued by employers as important for entry-level employees with and without disabilities. Career Dev. Transit. Except. Individ. **35**(1), 29–38 (2011). https://doi.org/10.1177/0885728811419167

I. Suarta, I. Suwintana, I. Sudhana, N. Hariyanti, *Employability Skills Required by the 21st Century Workplace: A Literature Review of Labor Market Demand* (2017). https://doi.org/10.2991/ictvt-17.2017.58

V. Thi, L. Dang, N. Khang, Twenty-first century essential employability skills for english as a foreign language undergraduates in a context of the Mekong Delta. Eur. J. Educ. Res. **11**(2), 1089–1102 (2022). https://doi.org/10.12973/eu-jer.11.2.1089

Chapter 9
How to Address Skills Gaps

A Review of Organizational Strategies

Manuel Souto-Otero

Abstract Closing skills gaps is a central problem confronting organizations. While much of the literature focuses on the quantification of existing skill gaps, the chapter reviews, instead, alternative approaches and strategies to deal with skills gaps. Three approaches are discussed: capability expansion—based on the development of new skills-, insourcing—based on the search for more efficient use of skills—and redesign—based on the reduction of the skills content of jobs. These approaches are distinct in terms of the strategies and actions that they require, timelines for implementation, and in their consequences for productivity, employment levels, and job quality.

Keywords Skill gaps · Skills matching · Skills development · Insourcing · Talent markets · Job redesign

9.1 Introduction

In human capital theory skills are presented as key factors for productivity and performance. This has led to a clear policy focus on skills supply as a route for economic growth. Authors like Hanushek (2017) have gone so far as to claim that for long-term economic development, only skills matter. In the short term, at least, this is clearly an oversimplified account of how productivity and competitiveness are achieved, as these depend on a wider set of environmental conditions (such as level of demand, regulatory environment, support infrastructures, nature of the competition, cost of skills), as well as on how work is designed, what skills are deployed at work, and how (Souto-Otero et al. 2023). Yet, there is a widespread agreement that skills are central to high performance.

Concerns over skills supply adopt many forms such as skills mismatches, skills shortages, and skills gaps (McGuinness et al. 2018). In this chapter, the focus is on

---

M. Souto-Otero (✉)
University of Bristol, Bristol, UK
e-mail: manuel.soutootero@bristol.ac.uk

S. Tillement et al. (eds.), *The Shaken Foundations of Safety*,
SpringerBriefs in Safety Management, https://doi.org/10.1007/978-3-032-22924-3_9

skills gaps, defined as situations in which workers do not possess adequate competences to perform their role proficiently (Pouliakas and Souto-Otero 2022), and more specifically on the strategies that organizations can adopt to address these. Some of these strategies are well-known, but the chapter also documents more novel strategies. Concerns with skills gaps, in both technical and soft skills (Cappelli 2015), have been growing worldwide (Rikala et al. 2024). Recent data shows that six in ten EU workers (63%) feel, to a great (16%) or moderate (47%) extent, that they need to develop their knowledge and skills to improve their job performance, and over half (52%) think that they need to develop their digital skills further to do their job better (Pouliakas and Souto-Otero 2022).

Skills gaps can expose workers and their organizations to a wide range of risks. These risks can include reductions in competitiveness and innovation, but also lack of compliance with existing regulations, or making unsound decisions in the absence of rules and regulations, including in the management of health and safety issues. It should be noted that not all gaps are negative. Skills gaps tend to emerge more frequently in dynamic work environments (Cedefop 2015) and more strongly affect employees in complex jobs. This is because gaps may be the result of organizational changes to introduce new products or new technology, which can be positive for organizations. They can also be the result of adaption to new regulations that improve quality or safety, or awareness of safer procedures and behaviors even in the absence of regulations. But even in such cases, organizations need to effectively respond to the emergence of gaps.

9.2 Information Gaps in Skills Gap Analysis

The identification of skills gaps is a first step to try and address them. However, human capital, unlike capital equipment, is not easy to see and measure. The topography of skills, moreover, is becoming increasingly complex. Commercial vendors on skills demand data captured through job advertisements now code tens of thousands of skills (Brown and Souto-Otero 2020). Those organizations that make efforts to identify skills gaps use various sources of data, forms of intelligence and forecasting to get a handle on present and future skills gaps. On the skills demand side, organizations use different forms of performance data, industry and occupational trends, employer surveys, job vacancy and job advert "big" data to map skills needs and identify gaps. However, technological advances, changes in consumer preferences and demands, or in the trade policy environment can quickly affect the accuracy of some of these tools.

Regarding the analysis of the supply side of skills, many organizations do not have a clear inventory of their stock of human capital—that is, the human capital of their workers. Organizations generally put great effort in assessing the skills of job candidates—through interviews, assessments and other techniques—but not in keeping an up to date overview of the skills of their existing workforce. A survey by McKinsey (2020) found that fewer than half of organizations have a clear view of their workers' skills. Under these conditions, it is difficult to optimize the matching between workers' skills and organizational skills needs. This results in greater potential for skills gaps in organizations—some of which may be perceived rather than genuine gaps, emerging from a lack of accurate and updated information on their workforce's skills.

Organizations can use various systems to make their stock of skills more transparent and identify gaps, such as performance reviews, skills inventories or skills audits—using tests, declarative methods (e.g., surveys, interviews), or observation (e.g., review of portfolios or samples from work). These tools help organizations to better understand workers' skills and facilitate their deployment on the right tasks. Organizations can also use proxies such as workers' stock of qualifications, including macro and micro-credentials, and professional certifications.

Digitalisation has provided new data sources for skills gap analysis. Many skills assessment and performance management tasks—including tracking of individual and team performance—are now facilitated by AI systems, rather than being conducted by HR or operations managers (Souto-Otero and Brown 2024). For example, professional networking social media platforms such as LinkedIn contain granular information on individuals' skills that organizations can scrape to improve matching, as well as to prevent gaps by monitoring industry trends. A frequent shortcoming is that skills gap analysis is often undertaken "one skill at a time" while organizations tend to require and recruit for bundles of skills. A trend of particular relevance here is the increasing importance of "hybrid" profiles that require specific combinations of both technical and soft skills (Brown and Souto-Otero 2020). This section has argued that skills gaps remain largely unknown in many organizations and underlined the importance of their identification. Once skills gaps are identified, they need to be addressed. The next section reviews various organizational approaches and strategies that can be adopted to deal with skills gaps.

9.3 Organizational Strategies to Address Skill Gaps

The literature on organizational strategies to address skills gaps is rather dispersed. This section discusses three main approaches (capability expansion, insourcing, and redesign) and related strategies to address skills gaps. The strategies that organizations can adopt depend on various factors, such as the kind of company they are—for example, in terms of their competitive strategies, size, or resources. In addition, technology is changing how organizations can respond to skills gaps, for example as the pool of workers that may be considered when a task needs to be completed can be expanded through the use of digital tools, that help locate the skills needed within various parts of the organization but also outside.

9.3.1 Capability Expansion: Increasing the Stock of Skills

Capabilities can be expanded through capability expansion: buying, building, or borrowing skills. A widely used strategy to reduce gaps is to increase the skills and capabilities of the pool of employees through "buying": recruiting new staff with the skills the organization needs—in some cases, an organization may "buy" (acquire or merge with) another organization, rather than hire individuals, to expand its skills capabilities. According to McKinsey (2020), recruitment is the most frequent strategy to combat skills gaps. Skills-based hiring, in particular, is seen as a way to deal with skills gaps (Fuller et al. 2022; Souto-Otero and Brown 2024). Recruiting can be more expensive than building or borrowing skills—see below—and there is a degree of uncertainty around how new recruits will perform, as skills are difficult to observe at the time of recruitment. When labor markets are tight and there is strong competition for workers with the skills employers want, they will need to compete with other employers in terms of the quality of the jobs that they offer—e.g., pay and working conditions, opportunities for flexible working, possibilities for skills and career development.

A second strategy to expand capabilities is building new skills through training and learning. This is the focus of much literature on skills gaps (Rikala et al. 2024), and can include the creation of sophisticated skills planning systems and learning journeys. Training can take multiple forms—internal or external, formal or non-formal (Souto-Otero 2021). Actions such as mentoring and job shadowing schemes are also part of this strategy, as is "learning by doing"—which is why workers on permanent contracts and greater job stability tend to report smaller skills gaps (Cedefop 2015). Actions to build new skills can be more long term than training the current workforce; they can be directed toward ensuring a future talent pipeline, for instance, through partnerships with education and training providers and contributions to their curricular design. The nature and effectiveness of training and upskilling actions are very diverse. For employers there can be direct monetary costs as well as opportunity costs associated with training actions. Some employers may also fear employee poaching when the training has been completed, although different financing mechanisms—such as employer levies—can help mitigate some of these issues (Souto-Otero 2007). Although in a context of flexible labor markets investment in employee training has lost some of its attraction, some forms of training such as job rotation, peer learning, and communities of practice can help limit the costs of building new skills.

A further strategy to expand capabilities is contracting (borrowing), to enlarge the pool of candidates considered for specific projects or tasks where the organization may have a skills gap, without hiring those candidates. An example of this is the construction of skill webs, whereby companies integrate a range of functions, strategies, and agents—such as its own workers, company suppliers or collaborators—based in different locations in their, sometimes global, sourcing of skills (Lauder et al. 2017). Candidates may be found through outsourcing including through platform work (Cullen and Farronato 2021). Borrowing has been greatly facilitated by the increasing use of collaborative technologies for work. The form of engagement with these new sources of skills varies significantly in its legal forms, frequency, or quality of engagement: from various types of gig work, to consultancy work and long-term partnerships.

9.3.2 *Insourcing: Increasing the Use of Skills*

Rather than developing or capturing new skills, organizations can resort to better use of the skills already available within them. This can be done through measures to make staff skills more transparent and to redeploy staff. This can take the form of redeployment into an existing role, flexibilization in the allocation of staff to tasks, or cross-functional collaboration. One example of this are Internal Talent Markets (ITMs), which are part of what has been called "skills-centerd organization." ITMs aim to reduce gaps by making employees' skills more transparent to enable skills deployment when needed, enhance employee motivation and retention, and reduce employees' downtime. This can be considered a form of "insourcing" (seeking the skills required within the organization), rather than outsourcing. ITMs bring skills demand into a single space, normally a digital platform. They are based on the fractionalisation of work—the breaking up of jobs and tasks into smaller, discrete units. ITMs tend to work better when organizations create not just competence frameworks for different roles, but a common skills taxonomy for the whole organization—which is then compatible with its HR and operations digital systems and can be used across functions such as recruitment, human capital management solutions, and learning experience platforms.

ITMs can function like social network sites where managers can browse profiles and place listings of tasks/ roles that need doing, and workers browse job offers. Both (hiring managers and workers) converge in the ITM space to find matches or let the ITM find the matches and notify them. In the latter case, candidates (workers) may create and update their profiles, browse jobs, and rank their preferred assignments. Algorithms then use these choices to suggest placements to workers. ITMs "force" employees to report on their recent skills development and make this visible to the organization, even when skills are not directly related to their current role, as they apply for new roles or tasks within the organization. Algorithms may also look for evidence of employees' skills—for example considering typical skills reported by people who had similar work, or outside work, experiences. Employees may need to self-assess their skills and profile in relation to the requirements in the openings available. In that process they identify gaps and training needs that they would need to close to opt for those roles that are of interest to them. They may also benefit from feedback from the managers who assess their applications. The idea is that agile forms of working based on task allocations—rather than fixed job role descriptions-, and cross-team work facilitated by ITMs can reduce skills gaps, as fixed job descriptions or team membership no longer prevent the organization from making use of the right skills in the right place and at the right time.

Novartis, for example, found through internal employee and exit surveys that in 2020/21 around 80% of its employees lacked visibility of career opportunities within the organization at large, and half were unaware of the skills that they needed to develop (Gloat 2022), information that ITMs can help them obtain. ITMs provide organizations with new internal talent and skills intelligence (who within the organization has skills that are needed in a given project and would be willing to participate in it) to inform their deployment and reduce gaps. Examples of other organizations that have experienced ITM include Schneider Electric or MasterCard.

While ITMs can help reduce skills gaps, they may also "displace" them, creating new gaps in some teams as they close them in others. They work better in organizations where skills requirements can change rapidly and where organizations have a large workforce and a good share of generalist profiles and tasks. Managers may be weary that ITMs lead to internal poaching of their best employees, and there is a risk of concentrating the most skilled employees in a small set of "desirable" teams. ITMs require investment from organizations, but above all a cultural change to think in terms of skills, more than fixed job roles and teams (Souto-Otero 2024).

Related to ITMs, but also more generally, staff motivation is a strategy that can reduce the presence of gaps derived from skills withdrawal. Negative reactions to job quality, for example, can lead disaffected individuals to withdraw in the use of their skills, generating a skills gap (Hurrell 2016), even if this could be considered a non-genuine gap because it is based on the purposeful non-deployment of employee skills.

9.3.3 Redesign: Reducing the Skills Content of Jobs

Organizations may reduce their skills gaps by reducing their need for skills through job (re)design or changes in their production strategies. Job (re)design and decisions around the bundle of knowledge, skills, competencies, and attitudes required "to do the job" can help reduce gaps. For example, work may become more specialized, fragmented, and "lean," to reduce skills gaps: instead of a large number of employees trying to be involved in all aspects of the production process they can be involved in specific stages and concentrate their skills in those parts of the process, following Taylorist principles. In such cases it is the team, not individual workers, who have the whole bundle of skills required for production.

Companies can also modify what they produce and the ways in which they produce to reduce gaps. Changes in product strategies can adopt various forms, such as greater specialization with regard to the range of products or services produced or reductions in the quality of what is produced, both of which can decrease the range or level of skills required during production processes.

Capital investment and automation strategies can also lead to the reduction of skills requirements. There is currently much uncertainty on the extent to which AI is leading or will lead to the replacement of workers, and—specifically—its potential to replace white collar workers, who were previously seen as the beneficiaries of technological developments (Frey and Osborne 2017; Souto-Otero et al. 2024). According to skills-biased technological change theory, capital investment in technology can lead to increases in skills requirements (as technology enhances the productive capacity of highly skilled workers) (Goldin and Katz 2008). But it can also result in reductions in the need for labor or in the skills that labor needs to possess, as suggested by Marx's (1976) discussion on deskilling, and captured more recently in the concept of digital

Taylorism—whereby knowledge work is transformed into working knowledge after it has been disaggregated into its constituent parts and codified (Brown et al. 2010). A final point to note is that digitalisation can increase skills gaps in the long term if it affects the talent pipeline. In those occupations where generative AI is capable of replacing those working in entry-level positions, this will affect the talent pool for senior roles in the future.

9.4 Conclusions

This chapter has reviewed the notion of skill gap, underlying that not all skill gaps are the result of negative dynamics, such as difficulties in attracting skilled staff, but can also be the by-product of positive dynamics. Innovative firms, for example, those that rapidly adapt to market demand conditions or modernize their technology more frequently, can be prone to experiencing gaps. But while skills gaps linked to rapid adaptation may signal dynamism, persistent gaps due to underinvestment in future-oriented skills and capabilities reflect organizational weaknesses.

In both cases, closing gaps is a central problem confronting organizations. While much of the literature has focused on the quantification of the level or extent of skill gaps, the chapter focused, instead, on alternative approaches and strategies to deal with them. The chapter has argued that while often the attention of policy-makers and discourses around skills gaps focus on the need to develop more skills, not all gaps can be solved by education, training, and skills supply measures. Organizations can consider a much wider set of strategies. The strategies discussed in this chapter are distinct in terms of the actions that they require, timelines for implementation, and in terms of their consequences for productivity, employment levels and job quality. One enduring challenge in the identification of skill gaps, and the adoption of effective solutions, is the absence of detailed and reliable data on the current stock of skills that organizations have—as well as lack of transparency regarding skills needs. This can result in information frictions and the conflation of genuine and perceived skills gaps, which in turn lead to the adoption of misplaced or ineffective strategies. While some innovative data sources to help address skill gaps have emerged, there is a need for much more thorough testing and refinement in their use. In the face of an uncertain future, anticipatory skill strategies are required to meet immediate skills gaps while also creating the basis for longer-term adaptability—helping to address gaps both the short and long term. Those firms that only aim to close current gaps and do not invest in those capacities needed to anticipate and respond rapidly to external changes risk being perpetually reactive, rather than capable of proactively managing their future. Similarly, firms with a degree of slack resources, in terms of skills availability, have a buffer against future shocks in a context of uncertainty. While this is likely to have some costs, these can be considered an investment in strategic agility to accelerate adjustment when new skill demands emerge.

References

P. Brown, H. Lauder, D. Ashton, *The Global Auction: The Broken Promises of Education, Jobs, and Incomes* (Oxford University Press, 2010)

P. Brown, M. Souto-Otero, The end of the credential society? An analysis of the relationship between education and the labour market using big data. J. Educ. Policy **35**(1), 95–118 (2020)

P.H. Cappelli, Skill gaps, skill shortages, and skill mismatches: evidence and arguments for the United States. ILR Rev. **68**(2), 251–290 (2015)

Cedefop, Skill shortages and gaps in European enterprises: striking a balance between vocational education and training and the labour market. Cedefop reference series; No 102 (Publications Office, Luxembourg, 2015).

Z. Cullen, C. Farronato, Outsourcing tasks online: matzching supply and demand on peer-to-peer internet platforms. Manag. Sci. **67**(7), 3985–4003 (2021)

C.B. Frey, M.A. Osborne, The future of employment: How susceptible are jobs to computerisation? Technol. Forecast. Soc. Chang. **114**, 254–280 (2017)

J. Fuller, C. Langer, M. Sigelman, Skills-based hiring is on the rise. Harv. Bus. Rev. **11**, 1–6 (2022)

Gloat (2022). https://gloat.com/gloat-live-2022/gloat-live-global-2022-watch-on-demand/?wchannelid=4hslktmpgh&wmediaid=amfppg7sgv. Accessed 07 July 2023

C.D. Goldin, L.F. Katz, *The Race Between Education and Technology* (Harvard University Press, 2008)

E. A. Hanushek, For long-term economic development, only skills matter. IZA World Labor 343 (2017)

S.A. Hurrell, Rethinking the soft skills deficit blame game: employers, skills withdrawal and the reporting of soft skills gaps. Hum. Relations **69**(3), 605–628 (2016)

H. Lauder, P. Brown, D. Ashton, Theorizing skill formation in the global economy, in *The Oxford Handbook of Skills and Training*. ed. by C. Warhurst, K. Mayhew, D. Finegold, J. Buchanan (Oxford University Press, Oxford, 2017)

K. Marx, *Capital*, vol. 1 (Penguin, London, 1976)

S. McGuinness, K. Pouliakas, P. Redmond, Skills mismatch: concepts, measurement and policy approaches. J. Econ. Surv. **32**(4), 985–1015 (2018)

McKinsey, Beyond hiring: how companies are reskilling to address talent gaps (2020). https://www.mckinsey.com/capabilities/people-and-organizational-performance/our-insights/beyond-hiring-how-companies-are-reskilling-to-address-talent-gaps. Accessed 15 Oct 2024

K. Pouliakas, M. Souto-Otero, Setting Europe on course for a human digital transition: new evidence from Cedefop's second European skills and jobs survey. Cedefop reference series; No 123 (Publications Office, Luxembourg, 2022)

P. Rikala, G. Braun, M. Järvinen, J. Stahre, R. Hämäläinen, Understanding and measuring skill gaps in Industry 4.0—a review. Technol. Forecast. Soc. Change **201**, 123206 (2024)

M. Souto-Otero, Access to post-compulsory education and training: economic, sociological and political determinants and remaining research gaps. Comp. Educ. **43**(4), 571–586 (2007)

M. Souto-Otero, Validation of non-formal and informal learning in formal education: covert and overt. Eur. J. Educ. **56**(3), 365–379 (2021)

M. Souto-Otero, European inventory on validation of non-formal and informal learning 2023 update: case study: Swiss Internal Talent Marketplaces (ITMs) Validation: the case of Novartis. European Commission and Cedefop (2024)

M. Souto-Otero, P. Brown, The rise of the digital labour market: characteristics and implications for the study of education, opportunity and work. J. Educ. Work **37**(1–4), 1:16 (2024)

M. Souto-Otero, P. Brown, S. Freebody, High skilled workplaces, technological change and employment: Can educational reform do it? Int. J. Educ. Res. **122**, 102265 (2023)

Chapter 10
Competencies and Skills for Navigating Uncertain and Interdependent Systems of the Future

Mark A. Griffin

Abstract Organizations must navigate an external environment where new opportunities continuously generate new risks to be managed. A cycle of disruptive innovation followed by regulatory constraint seems to be occurring more rapidly and more dramatically. This chapter explores the way this cycle in the external environments shapes the need for both innovation and control in all work systems including those that manage productivity and safety. To adapt organizations must not only balance these competing demands but prioritize strategic orientations over time. Frameworks from individual role performance, organizational dynamic capability, and uncertainty management are combined to identify the skills and competencies most needed in this environment.

Keywords Uncertainty · Work systems · Dynamic capability

10.1 Introduction

The industrial world is facing a convergence of rising uncertainty and deepening interdependence in the twenty-first century. *Uncertainty* refers to a lack of predictability in the external environment, while *interdependence* describes how different parts of a system are tightly connected, creating complexity and a need for coordination. Over the past few decades, technological advances—from global communication networks to automated decision systems—have massively increased these connections and dependencies, making outcomes harder to predict. Earlier assumptions of the late twentieth century held that change was constant yet relatively steady and manageable, but the last 15 years have challenged this view. Instead of linear change, we have witnessed dramatic swings between periods that embrace uncertainty (encouraging innovation and open systems) and periods that control uncertainty (tightening rules and boundaries).

M. A. Griffin (✉)
Curtin University, Perth, Australia
e-mail: mark.griffin@curtin.edu.au

S. Tillement et al. (eds.), *The Shaken Foundations of Safety*,
SpringerBriefs in Safety Management, https://doi.org/10.1007/978-3-032-22924-3_10

10.2 Fifteen Years of Uncertainty and Interdependence (2009–2024)

10.2.1 Global Financial Crisis (2008–09)

The 2008 financial crisis was a shock that revealed hidden interdependencies in the world economy. In the years leading up to 2009, an appetite for risk and financial innovation created a complex web of mortgages, securities, and leverage that seemed to yield predictable profits—until it collapsed. When the U.S. housing bubble burst, highly leveraged banks around the globe faltered, triggering a cascade of uncertainty: credit markets froze, unemployment spiked, and economic instability prevailed. This disruption highlighted how tightly connected financial instruments and institutions had become, and how quickly a local problem could amplify into global turmoil. In response, regulators swung the pendulum toward control. Major reforms—such as the U.S. Dodd-Frank Act (2010) and Europe's new supervisory mechanisms—imposed stricter controls to reassert stability. These measures aimed to rebuild confidence by curbing risky behaviors and increasing transparency. While heavy regulation reduced some uncertainties in financial systems, it also introduced complexity and compliance costs, raising questions about unintended consequences such as driving risk into "shadow banking" sectors (Xie 2022). The financial crisis thus exemplified a cycle: high levels of innovation and interdependence reaching a breaking point, which then prompted an era of tightening and oversight to increase control.

10.2.2 Global Pandemic (2019–21)

A decade later, the COVID-19 pandemic again created extreme uncertainty on a global scale. The emergence of a novel coronavirus at the end of 2019 meant that much about the threat was unknown—how it spread, how to treat it, or its long-term impacts. This scientific and medical uncertainty, coupled with fear, led to wide-ranging economic shocks and social disruption. Crucially, the pandemic exposed the vulnerabilities of global interdependence: international supply chains that normally functioned were suddenly paralyzed as factories shut down and cross-border flows stalled (Xu et al. 2020). Shortages of critical goods illustrated how tightly coupled and fragile the "Just-in-Time" global production networks had become. In the face of this instability, governments around the world reacted by reasserting control through unprecedented regulations on daily life. Lockdowns, travel bans, and emergency public-health mandates swept across countries in an effort to contain the virus and stabilize society. These strict measures—while understandable for managing the immediate crisis—also had the effect of pulling back from global interdependence. Many nations reevaluated reliance on distant suppliers and began to emphasize domestic resilience over efficiency. The pandemic's aftermath has been characterized by calls to shorten supply chains and build buffers, representing another swing toward closing off uncertainty after a period of global openness.

10.2.3 Generative AI Revolution (2020s)

As the world emerged from the pandemic, a new disruption gained momentum: the rapid rise of generative artificial intelligence. Recent advances such as large language models and highly autonomous systems are creating vast opportunities alongside new uncertainties. On one hand, these technologies can absorb massive, interconnected data and produce human-like insights, promising breakthroughs in productivity and innovation. On the other hand, their novelty and speed of development make it impossible to predict how they will evolve or what consequences they will bring. This fuels a sense of uncertainty across industries—firms cannot be sure which skills will be automated next or what new business models might emerge. At the same time, AI systems and digital platforms are connecting the globe even closer together through data-sharing and algorithmic decisions; all increasing interdependence (Remolina 2023). The rapid development of generative AI applications since 2023 is multiplying opportunities and risks (Abudawood 2023). While generative AI is increasing all aspects of connectivity, the associated uncertainty is surpassing all previous levels. The nature work is expected to change in ways that will reverberate through all levels of the economy. Regulators, too, are grappling with how to encourage innovation in AI while preventing harm. For example, analysts note that authorities must simultaneously incentivize beneficial AI development and constrain dangerous applications despite great uncertainty about these technologies' risks (Bryan and Teodoridis 2024).

10.3 Frameworks for Understanding Responses to Uncertainty

To navigate high uncertainty and interdependence, organizations need to manage both human behavior and strategic capabilities. We outline three complementary frameworks: (a) a model of individual work role performance under uncertain, interdependent conditions, (b) the dynamic capability framework from strategic management, and (c) a model of uncertainty management through opening and closing phases.

10.3.1 Individual Work Role Performance in Uncertain, Interdependent Contexts

When environments are unstable and complex, what employee behaviors help organizations succeed? Griffin et al. (2007) propose that in contexts of uncertainty and interdependence, traditional job proficiency must be coupled with adaptivity and proactivity, as well as behaviors that support coordination with others. **Adaptivity** is the ability to adjust one's actions and learn new ways of working in response to unexpected changes. For example, adaptable employees readily update their procedures and roles when technologies or regulations change, helping maintain safe and effective operations even as conditions shift. **Proactivity** is the tendency to take initiative by anticipating future challenges or opportunities and acting in advance. A proactive worker might, for instance, report a near-miss and suggest a preventive solution before an accident happens, or propose an innovative workflow to improve efficiency. Both adaptivity and proactivity are forward-looking, change-oriented behaviors that enable an organization to stay ahead of crises rather than merely react. These qualities are especially valuable when threats and opportunities cannot be fully predicted in formal plans.

In highly interdependent systems, individual success also depends on collaboration and coordination with others. No single employee can have all the knowledge or control in a complex operation, so people must work together across teams and specialties. **Collaborative** behavior means sharing information, assisting peers, and jointly problem-solving, which ensures that insights and warnings travel quickly through the network. **Coordinated** action means aligning one's tasks with the broader workflow—for example, following agreed protocols during shift handovers or synchronizing steps in a supply chain. Effective coordination provides stability: even when individuals are adapting rapidly, they do so in a way that keeps the overall system running smoothly rather than chaotically.

10.3.2 Dynamic Capabilities for Strategic Adaptation

At the organizational level, the dynamic capability framework explains a process of managing the competing needs to explore the changing environment for new possibilities and exploit the resources available in a stable environment (O'Reilly and Tushman 2004). David Teece and colleagues define dynamic capabilities as a company's ability to *sense* changes in the environment, *seize* opportunities, and *transform* itself accordingly (Teece et al. 1997). *Sensing* involves continuously scanning for external opportunities and threats by gathering market intelligence, detecting weak signals such as emerging technologies or shifting customer preferences, and challenging internal assumptions. Once opportunities are identified, *seizing* requires mobilizing resources quickly—making fast decisions, allocating capital, and executing initiatives like launching new products or forming strategic partnerships, even amid uncertainty. Finally, *transforming* refers to the ongoing reconfiguration of organizational structures, skills, and processes to sustain adaptability. Based on the above principles, Griffin et al. (2016) introduced the concept of "dynamic safety capability," which refers to a company's capacity to continually update and improve its safety systems in anticipation of emerging hazards. A firm with strong dynamic safety capability constantly learns from minor incidents, adapts its safety protocols before major accidents occur, and encourages front-line employees to suggest safety innovations. Managing safe operations in turbulent times requires a dynamic approach to safety that goes beyond managing stability to include exploration of change.

10.3.3 Uncertainty Regulation by Phase: Opening and Closing Cycles

While the first two frameworks focus on specific levels (individual and firm), the third framework by Griffin and Grote (2020) offers a multi-level perspective on how organizations regulate uncertainty through cycles of "opening" and "closing" actions. These are two fundamentally different modes of responding to uncertainty, and effective systems cycle between them over time. In an opening phase, organizations and individuals embrace uncertainty: they encourage flexibility, exploration, and the generation of new possibilities. This phase is characterized by *adaptive* and *proactive* behaviors—employees are empowered to adjust their roles, experiment with novel solutions, and pursue opportunities without strict permission hierarchies. Conversely, in a closing phase, the emphasis shifts to controlling uncertainty by tightening structures and reinforcing what is known. Here the organization prioritizes stability, efficiency, and risk mitigation. Roles and processes become more formalized; employees are expected to focus on proficiency—executing defined tasks with consistency and discipline—rather than on improvisation. Crucially, both modes are necessary. Organizations must dynamically cycle between opening and closing to match the environment. This cycling can occur at different paces—sometimes within project life cycles or management routines—and at different levels.

To implement these cycles, individual competencies and career paths will develop in new ways. As generative AI supports or replaces diverse cognitive tasks, the need for critical thinking will, perhaps counterintuitively, become a more important skill. Individuals will apply information processing skills to evaluate more diverse and variable sources of information. Information processing skills will also be required in a context of continuous learning where individuals must learn and apply new information quickly. Career paths within specific disciplines such as healthcare will require integration of AI tools with human knowledge of insight. Some careers will depend less on a defined domain of knowledge but on the ability to integrate new domains of knowledge. A strategic orientation directed toward designing change and training others will be critical.

10.4 Integrating the Frameworks: Managing High Uncertainty in a Connected World

Taken together, the three frameworks highlight the need for effective organizations to be ambidextrous—to cultivate both the capacity to flexibly explore and the capacity to efficiently exploit, and to know when to shift focus. Managing high uncertainty in a globally connected context cannot be a matter of choosing one approach but of integrating these levels of response as the external environment demands relatively rapid shifts between an innovation focus and a regulation focus (Muruganandan et al. 2022). The table below summarizes how the three frameworks can be integrated and provides strategic guidance for engaging with uncertainty and interdependence (Table 10.1).

10.4.1 Align Individual Behaviors with Strategy

The proactive and adaptive work behaviors from the first framework provide the human foundation for dynamic capabilities. For example, a company's ability to sense opportunities depends on employees across levels actively scanning the environment and sharing information. Its ability to seize opportunities relies on individuals being empowered to make changes and on cross-functional teams coordinating swiftly. In stable times, an emphasis on proficiency and coordination supports the transforming capability by embedding reliable routines. Thus, dynamic capabilities (sensing, seizing, transforming) are most effective when the workforce possesses the corresponding competencies to execute them in either an innovative (opening) or disciplined (closing) manner.

10.4.2 Phase Shifts Between Opening and Closing Orientation

Knowing when to shift is also important (Papachroni et al. 2015). Leaders play a critical role in monitoring the external environment and internal performance to decide if an opening orientation or a closing orientation should dominate at a given time. Leaders can establish sensing mechanisms—such as environmental scanning teams, risk dashboards, and feedback loops—that alert the organization to rising uncertainty or emerging stability. For example, a sudden change in customer demand or a geopolitical shock might trigger a planned pivot to more exploratory, adaptive operations (opening). Conversely, signs that variance in outcomes is increasing beyond acceptable levels (e.g., quality issues, safety incidents) might signal a need to tighten controls (closing).

Table 10.1 Integration of three complementary frameworks navigating uncertainty

Dynamic capability	Opening orientation adaptive and proactive behavior to engage uncertainty	Closing orientation proficient and structured behavior to manage uncertainty
Sensing *Identifying opportunities and threats*	Emphasize exploratory sensing. Employees engage in **proactive** environmental scanning and external research beyond routine channels, seeking new information and weak signals early. They show **adaptivity** in adjusting what they monitor as trends change, often **collaborating** across units or with external partners to gather diverse insights. This open approach increases flexibility to detect emerging opportunities under uncertainty	Emphasize systematic sensing. Employees rely on **proficient** data-gathering and analysis using established indicators and processes. Sensing is conducted in a **structured** manner—following set routines, checklists, and known information sources. **Coordination** of information flows (e.g., regular reports, dashboards) ensures everyone focuses on the same key metrics. This closed approach reduces uncertainty by focusing on well-understood market signals and proven analysis methods
Seizing *Capturing opportunities and designing responses*	Emphasize agile seizing of opportunities. The organization encourages **proactivity** in idea generation and quick decision-making to capitalize on new opportunities. Teams **adapt** by pivoting resources or business models as needed, and **collaboration** is high (cross-functional teams brainstorm and experiment) to innovate solutions. Opening behaviors here mean flexibly **seizing** nascent opportunities under uncertain conditions (tolerating some risk and iteration)	Emphasize efficient exploitation of opportunities. The organization focuses on **proficient** execution of chosen strategies—refining the business case, planning, and implementing with discipline. **Structured** processes (e.g., stage-gate reviews, project management frameworks) and **coordination** of resources (clear roles, timelines) are used to roll out the opportunity. Closing behaviors ensure the seized opportunity is integrated predictably and profitably, fitting known structures and reducing risk
Transforming *Continuously renewing and reconfiguring*	Emphasize adaptive transformation. Leaders and employees **proactively** drive change initiatives (seeking improvements or new directions even before crises hit). The organization shows **adaptivity** by flexibly reconfiguring structures, processes, and resources to adjust to emerging environments. Widespread **collaboration** and learning are encouraged (sharing knowledge, involving teams in change design) to build buy-in for transformation. This opening approach enables constant renewal under high uncertainty, fostering innovation and agility	Emphasize incremental, well-managed change. The organization **proficiently** implements transformations through careful planning and execution—ensuring new processes or structures work reliably. Changes are institutionalized via **structured** change management: formal training, standard operating procedures, and **coordination** across departments to align the transformation with existing operations. This closing approach focuses on stability and efficiency, suitable for lower-uncertainty periods where major shifts are less frequent

These iterations have direct implications for the skills and competencies across an organization. Opening periods require people to have skills in agile thinking and decision-making with the ability to generate and evaluate future scenarios. As AI increases, people will move from production-oriented roles to those involving judgement and creativity. These skills will be critical for exploring dynamic environments. Organizations will need to nurture these skills and support learning during opening periods. During periods of greater control, organizations will value skills of vigilance and ethical awareness. Resilience during these times will also require complex thinking but with a greater attention to detail and risk scenarios.

In essence, organizations must manage the skills and careers of its people to move *back and forth* between two complementary states. Research on high-performing firms underscores that those able to "toggle" their dominant orientation as needed—ramping up exploration when facing disruption and refocusing on exploitation when conditions stabilize—achieve greater resilience and performance.

10.4.3 Developing Dynamic Safety Capability

Dynamic safety capability can be viewed as the integration of the above two frameworks to enable safety in a continually changing environment. It is reflected in continual learning and controlled risk-taking to ensure ongoing adaptivity. Encouraging safe experimentation (a hallmark of opening behavior) can help discover solutions for unprecedented problems, but it should be coupled with mechanisms to contain potential harm (a function of closing behavior). For example, companies can use "sandboxes" or trial programs to test innovations on a small scale, which allows learning under uncertainty without endangering the whole operation. Likewise, maintaining slack resources or backup systems (redundancy) provides a safety net during opening phases when variability is higher. Over time, each cycle of opening and closing should grow the organization's knowledge base: lessons from experiments and crises are captured and embedded into standard practices or training, enhancing the firm's dynamic safety capability. By doing so, even regulatory compliance and safety protocols become more adaptive—shifting from a rules-based focus to a more proactive, risk-based approach that can handle novel situations.

The human skills and competencies required to build dynamic safety capability must support opening and closing strategies and also help to balance and integrate these strategies. Skills related to opening include scanning for emerging issues, flexible thinking, and continuous learning and experimenting. Closing strategies are supported by risk awareness, conscientious vigilance, critical evaluation, and psychological readiness. Integration requires skills in recognizing the strategy that is prevalent, evaluating trade-offs between safety and innovation, and building trust through communication.

10.5 Conclusion

The coming decades promise rapid technological change in a context of political instability, climate change, and pandemics, which form a volatile backdrop for all enterprises (World Economic Forum 2024). The interplay of opening and closing strategies in organizations will continue to demand new skills and competencies from individuals to ensure organizations exhibit dynamic safety capabilities. Across opening and closing strategies, the ability to adapt to change and integrate new information will be especially important. The increasing use of AI will not replace these skills, but the knowledge to be integrated will constantly evolve. Continuous learning, therefore, will be an intrinsic feature of skill development and career opportunities with greater requirements for adapting to changing environments. Safety will depend on scanning the horizon for emerging threats and opportunities originating anywhere in the world.

References

T.A. Abudawood, What cybersecurity threats does generative AI expose us to? World Economic Forum (2023)

K. Bryan, F. Teodoridis, *Balancing Market Innovation Incentives and Regulation in AI: Challenges and Opportunities* (Brookings Institution, 2024). https://coilink.org/20.500.12592/2hn2fw6

M.A. Griffin, G. Grote, When is more uncertainty better? A model of uncertainty regulation and effectiveness. Acad. Manag. Rev. **45**(4), 745–765 (2020)

M.A. Griffin, J. Cordery, C. Soo, Dynamic safety capability: how organisations proactively change core safety systems. Organ. Psychol. Rev. **6**(3), 248–272 (2016)

M.A. Griffin, A. Neal, S.K. Parker, A new model of work role performance: positive behaviour in uncertain and interdependent contexts. Acad. Manag. J. **50**(2), 327–347 (2007)

K. Muruganandan, A. Davies, J. Denicol, J. Whyte, The dynamics of systems integration: Balancing stability and change on London's Crossrail project. Int. J. Project Manag. **40**(6), 608–623 (2022)

C.A. O'Reilly, M.L. Tushman, The ambidextrous organisation. Harvard Bus. Rev. **82**(4), 74–83 (2004)

A. Papachroni, L. Heracleous, S. Paroutis, Organizational ambidexterity through the lens of paradox theory: Building a novel research agenda. J. Appl. Behav. Sci. **51**(1), 71–93 (2015)

N. Remolina, Interconnectedness and financial stability in the era of artificial intelligence, in *Artificial Intelligence in Finance* (Edward Elgar Publishing, 2023), pp. 350–367.

D.J. Teece, G. Pisano, A. Shuen, Dynamic capabilities and strategic management. Strat. Manag. J. **18**(7), 509–533 (1997)

World Economic Forum, The global risks report (2024). https://www.weforum.org/publications/global-risks-report-2024/in-full/

Z. Xie, *The Unintended Consequences of Banking Regulations: Shadow Banking* (Regulatory Studies Center, George Washington University, 2022)

Z. Xu, A. Elomri, L. Kerbache, A. El Omri, Impacts of COVID-19 on global supply chains: facts and perspectives. IEEE Eng. Manag. Rev. **48**(3), 153–166 (2020)

Chapter 11
Meeting the Skills Challenge

Fostering a Compensation Strategy for Safe Performance

Stéphanie Tillement

Abstract This concluding chapter revisits the principal contributions of the volume and seeks to deepen their analysis through the lens of the strategies that organizations may deploy to address the challenge of maintaining and developing skills under conditions of uncertainty and pervasive scarcity. Three strategies are discussed: adjustment, restoration, and compensation. Compensation appears as the most promising, but also the most demanding strategy. A central question frames this discussion: who assumes responsibility for organizing and equipping skills maintenance and development so as to sustain safe performance?

Keywords Adjustment · Restoration · Compensation · Scarcity · Responsibility · Skills distribution · Safe performance

11.1 Introduction

This book opens with a fundamental observation: safety is part of the job. In other words, safe performance is inextricably linked to workers' practices and skills, from top to bottom. The central question of the book follows: how can we avoid the erosion of safety due to skills shortages and gaps?

The chapters in this volume point to a central insight: skills are neither static nor merely individual attributes; they are dynamic, situated, and deeply embedded in social, organizational, and technological contexts. This perspective challenges traditional approaches that treat skills as fixed resources to be acquired and stored. Instead, skills emerge through interaction—between people, technologies, and institutions—and evolve across time and space, from local work situations to inter-organizational trajectories and ecosystems.

S. Tillement (✉)
IMT Atlantique, RESOH Chair, LEMNA, Nantes, France
e-mail: stephanie.tillement@imt-atlantique.fr

S. Tillement et al. (eds.), *The Shaken Foundations of Safety*,
SpringerBriefs in Safety Management, https://doi.org/10.1007/978-3-032-22924-3_11

Across the three parts of the book, a common thread is the recognition that safety and performance cannot be disentangled from the ways organizations anticipate, develop, and sustain skills. Part I highlights the structural tensions inherent in skills management: the coexistence of individualized career paths and collective work realities, the fragility of skill ecosystems under discontinuous industrial activity, and the paradoxical effects of digitalization, which can simultaneously enable learning and induce deskilling.

This volume, particularly through the analyses and solutions presented in Parts II and III, demonstrates that many organizations—especially those operating in high-risk contexts—are fully aware of the strategic importance of skills and are actively intensifying their thinking and initiatives in this domain. At the same time, the chapters bring to light a persistent paradox. Despite the availability of increasingly powerful tools and innovations for the collection, processing, and dissemination of data and knowledge, skills management remains a significant organizational challenge. This observation invites a dual line of inquiry. On the one hand, it raises fundamental questions concerning the nature of skills themselves, an issue that is examined in depth throughout the book. On the other hand, it compels closer scrutiny of the organizational work through which skills are structured, sustained, and made effective, whether voluntarily or not. This organizational work cannot be assumed to be spontaneous, natural, or transparent; rather, it is the outcome of deliberate, and often contested, processes of design, coordination, and governance.

Documenting the ways in which organizations and their stakeholders confront skills fragility, and consequently organize and equip skills management, thus emerges as a central analytical challenge. Across the chapters of this volume, three main categories of actors are repeatedly identified as playing a decisive role in these processes: managerial and strategic actors responsible for setting overall directions and allocating resources; human resource actors who design, formalize, and operationalize skills management devices and standards; and operational actors who mobilize, develop, and sustain skills in situated work activities. These actors intervene through a combination of formal mandates and informal practices, whether by choice or by necessity. However, the expectations placed upon each of these groups and, more importantly, the ways in which their respective contributions are articulated, coordinated, and governed are still somewhat of a mystery. Simply put, *everybody is involved in skills management, but who is steering it*? In high-risk organizational contexts, this uncertainty is further compounded by the presence of a fourth set of actors—those responsible for safety—whose positioning within, and relationship to, the three aforementioned categories warrant closer examination. This is the path we are now exploring.

11.2 Managing and Sustaining Skills: Who's in Charge?

In 1911, Frederic Taylor, known as the father of the Scientific Organization of Work, wrote: "*The search for better, for more competent men (sic), from the presidents of our great companies down to our household servants, was never more vigorous than it is now. And more than ever before is the demand for competent men in excess of the supply. What we are all looking for, however, is the ready-made, competent man; the man whom someone else has trained. It is only when we fully realize that our duty, as well as our opportunity, lies in systematically co-operating to train and to make this competent man, instead of hunting for a man whom someone else has trained, that we shall be on the road to national efficiency*" (Taylor 1911: 6).

Taylor's approach contains elements that may be regarded as surprisingly modern, particularly in his conception of skills not as attributes external to the individual, the job, or the organization, but rather as constructs that are produced within and by the organization itself. From this perspective, skills must be deliberately designed, structured, and equipped by the organization. One of Taylor's principal contributions lies in his explicit emphasis on organizational responsibility for skills management and in his direct questioning of who should assume responsibility for skills development. In Taylor's framework, responsibility is removed from workers and line managers and reassigned to executives, who institutionalize it within a technostructure operating outside the direct line of command. His approach thus advocates a highly centralized and formalized approach to skills management, governed by explicit rules that, in his view, should not be entrusted to workers themselves. This perspective was subsequently subject to strong criticism, particularly on the grounds that it contributed to the deskilling of workers. Despite these criticisms, Taylor's work raises a fundamental and enduring question regarding responsibility for skills management: who's in charge? Who should be in charge?

This crucial question ultimately runs through the entire volume, albeit often implicitly. This issue is all the more pressing given the contemporary developments that have profoundly transformed the context in which skills are conceptualized and managed. Powerful megatrends are reshaping the skills landscape. Among these, artificial intelligence is fostering increasingly complex forms of hybridization between human and non-human actors in the workplace. In addition, environmental, demographic, and broader societal transformations—such as declining sectoral attractiveness, image deficits in certain industries, and evolving relationships to work and organizational commitment—are exerting a significant influence on both the nature of skills and the ways in which they are managed.

Just as it is now widely acknowledged that there is no "one best way" of organizing work, there is no single or definitive answer to the question of who is responsible for skills management. One reason for this indeterminacy is that, despite its strategic importance, this question frequently remains what many industrial and safety practitioners describe as an "orphan issue": a domain that attracts limited attention and few dedicated resources within organizations.

With regard to safety professionals in particular, this concern often lies well beyond their formal authority, however regrettable this may be, both for them and for safety outcomes. In many cases, the allocation of responsibility is not only outside their control but also beyond their sphere of influence. Safety managers are frequently criticized for relying on standardized safety training; yet, in numerous organizational contexts, their capacity to act is constrained by the limits of their mandate. This situation reflects a broader paradox: if safety is understood as integral to work activity, then safety-relevant skills necessarily extend far beyond the exclusive remit of safety specialists. As a result, safety managers remain dependent on the willingness of more powerful actors—most notably senior management—to recognize ownership of the issue and to involve them meaningfully in skills-related decisions.

In contexts of structural scarcity, the question of who is in charge of skills management, and in what ways, is shaped by the strategies organizations adopt to address skills shortages, as well as by a range of contextual factors, including sectoral characteristics, technological configurations, and organizational arrangements.

11.3 Performing as Well with Less: Adjusting, Restoring, or Compensating

It is now a universally recognized fact: organizations are increasingly confronted with the renewed risk of shortages in critical resources and competencies, and are compelled to maintain—if not improve—levels of performance while operating with fewer resources, or even in conditions where these are partially absent. This situation foregrounds a fundamental analytical question: what strategies do organizations currently deploy, or could they deploy, to address these constraints on skills and resources? With what effects on safety?

Drawing on the chapters assembled in this volume, as well as on extensive exchanges with practitioners from high-hazard industries, we identify three dominant strategic orientations, examined in detail below. Each constitutes a distinct response to skills shortages and gaps, while simultaneously bringing to the fore a fundamental—yet often overlooked—question for organizations: who assumes responsibility for the management, maintenance, and development of skills?

11.3.1 Adjustment: Tacit Acceptance and the Organizational Production of "Misery"

The first response to skills scarcity may be characterized as **adjustment**, understood here as a strategy of limited or absent organizational intervention. Simply put, the organization "chooses" to do nothing, or very little. In this configuration, skills shortages are either not explicitly recognized or are treated as unavoidable constraints to which the organization will adapt incrementally, by tinkering and improvising locally. Such an approach may reflect blindness, weak anticipatory capacities, decision-making inertia, or an implicit belief in the organization's ability to cope through informal adaptation and individual resilience. Some industries, such as construction, have developed an adaptive culture, succinctly captured by the motto, "We will find ways, as we always do." More deliberately, organizations may also assume that existing margins or organizational slack allow them to absorb the effects of scarcity, and tacitly accept the eventual costs in terms of degraded safety, technical reliability, or economic performance.

From the perspective of safety studies, adjustment entails a tacit redefinition of acceptable risk. By tolerating declining margins, organizations implicitly normalize deviations from nominal operating conditions. The resulting trade-offs are rarely made explicit, yet they redistribute risk unevenly across stakeholders. Front-line operators, who are directly exposed to hazardous situations, are required to compensate for organizational shortcomings through local adaptations. Over time, this dynamic contributes to a process of "normalization of misery," in which degraded working conditions and elevated risk become routinized. Management's role in this process is less one of active decision-making than of implicit consent, as the burden of coping with scarcity is effectively transferred to operational staff. Safety managers and safety experts are expected to find local solutions and help operational staff adapt to constraints.

11.3.2 Restoration: Replenishment Through Planning and Standardization

The second response, which we refer to as restoration, is grounded in the explicit recognition of skills scarcity and in the organizational ambition to return to a predefined reference level of resources and competencies. This strategy is closely aligned with traditional HRM approaches and relies on forecasting, workforce planning, and standardized training mechanisms. Through long-term projections, organizations seek to identify quantitative deficits and, to a more limited extent, qualitative needs in terms of generic skills. Corrective measures are then deployed to rebuild capacity, including recruitment campaigns, internal training programs, and sector-level educational infrastructures, such as corporate or industry-specific universities. In this context, safety managers and safety experts are expected to double their efforts, but no significant change is required in their mode of action.

In safety and organizational terms, restoration aims to re-establish lost safety margins by replenishing human and technical resources. While often indispensable, this approach exhibits structural limitations. Its reliance on long temporal horizons makes it poorly suited to situations of acute or rapidly evolving scarcity. Moreover, restoration tends to privilege codified and transferable skills, which are easier to identify, measure, and train. As a result, it frequently struggles to capture and regenerate more situated, experience-based, or collective competencies that are critical to safe and reliable operations in complex socio-technical systems. The nuclear industry's demand for welders provides a particularly instructive example. Given the critical role of welding in the safe operation of nuclear systems, nuclear welders cannot be treated as interchangeable with welders operating in other industrial settings. Their work requires a high level of specialized know-how and technical expertise, which can only be developed through extended experience and sustained practice. As a result, generic welding competencies must be "situated" within the specific technical, organizational, and safety constraints of the nuclear domain. It is precisely in this necessary "situating" work that restoration strategies often encounter their limits.

11.3.3 Compensation: Organizational Reconfiguration and the Redistribution of Skills

The third strategy, which remains largely underdeveloped in both practice and theory, is what we refer to as compensation. Like restoration, compensation starts from an explicit acknowledgment of scarcity. However, it departs from a logic of replenishment and instead adopts a logic of organizational reconfiguration that is driven by a redistribution of skills where they are most needed to perform safely. The central premise is that organizations already possess a significant stock of skills and knowledge, but that these resources are imperfectly mobilized, coordinated, or aligned with operational demands.

From this perspective, skills are understood not merely as individual attributes but as emergent properties of work systems, shaped by task design, coordination mechanisms, and learning processes. Compensation therefore consists in rethinking how work is organized in order to enhance the effective use of existing competencies. This may involve redesigning workflows, adjusting interfaces between functions, fostering cross-boundary cooperation, and creating conditions for collective and situated learning. Rather than seeking to restore an idealized past state, compensation aims to sustain performance and safety under conditions of persistent constraint.

Implementing such a strategy requires profound organizational change and challenges established distributions of authority and responsibility. It demands the active involvement of multiple actors and a shift away from siloed approaches to skills management. Human resource functions, while central, must move beyond macro-level, individualized frameworks such as conventional forward-looking management of jobs and skills (GPEC), and engage more directly with operational contexts. At the same time, operational actors must be granted the temporal and organizational resources necessary to step back from short-term performance pressures and contribute to longer-term skill development through collective practices. Decision-makers and managers have a key role to play: as Perrow (1994) advocated years ago, this entails bringing the system's elites closer to operational practices and associated risks, thereby enabling them to develop a contextualized understanding of the critical skills underpinning safe production and to commit actively to their recognition, support, and effective mobilization. Safety managers and safety experts need to be actively involved across the entire process, from planning to implementation. Experts in human and organizational factors of safety (HOFS) are especially well positioned to contribute, in light of their competencies in activity and organizational analysis related to safety issues.

At present, both the governance of this convergence between actors and the concrete organizational arrangements that would support it remain largely to be invented. As such, compensation represents not only a strategic option but also a research agenda at the intersection of safety science, organizational design, and skills management (Table 11.1).

11.4 To Conclude... Provisionally

In times of scarcity, whether of resources or skills, organizations must adapt to gaps and uncertainty, finding ways to maintain or improve performance with less. Among the strategic approaches discussed, the third, *compensation,* is particularly promising. Implementing it has implications not only for how skills are conceptualized, but also for how resources and competencies are collectively mobilized and organized.

Key principles for such a strategy include:

- focusing on skills in context rather than as generic attributes;
- promoting circulation, sharing, and transmission of skills through mentoring or tutoring;
- recognizing and reinforcing the collective dimension of skills;
- engaging internal stakeholders and coordinating their cooperation;
- moving beyond organizational silos to enable dialogue;
- encouraging innovation in technology, work organization, and training;
- fostering collaboration between human resources, operational units, and safety experts—including specialists in human and organizational factors—to support "skillful and safe performance."

More broadly, compensation requires reconstructing a negotiated order (Schulman 1993) of safe performance. This entails the establishment of transparent and sustained dialogue among stakeholders, centered on the redefinition of performance objectives, an essential condition for achieving equal or improved outcomes under conditions of constraint. It is inherently an organizational matter, one that cannot be separated from the growing complexity of contemporary organizations. This complicates resource allocation, responsibility assignment, and learning. Ultimately, adopting a compensation strategy also entails the demanding work of simplification... but that is a topic for another discussion.

Table 11.1 Comparative framework of organizational strategies in the face of skills scarcity

Analytical dimension	Adjustment	Restoration	Compensation
Recognition of scarcity	Implicit, denied, or naturalized; scarcity treated as an external constraint	Explicit recognition of shortages relative to a reference level	Explicit recognition of persistent scarcity as a structural condition
Strategic intent	Cope with constraints through tolerance and informal adaptation	Replenish resources and skills to restore a predefined standard	Sustain or improve performance and safety through organizational reconfiguration
Temporal horizon	Short term, reactive	Medium to long term, anticipatory	Medium term, iterative, and adaptive
Locus of control	Decentralized in practice but ungoverned; burden shifted to front-line actors	Centralized and formalized (HR and management-led)	Distributed and coordinated across organizational levels
Conception of skills	Individual and taken-for-granted; skills assumed to "stretch"	Codified, generic, and transferable skills	Situated, collective, and system-level competencies
Primary mechanisms	Informal workarounds, tolerance of degraded conditions	Workforce planning, recruitment, training programs, sectoral education	Work redesign, coordination mechanisms, collective learning, skill redeployment
Impact on front-line work	Increased workload, exposure, and responsibility without formal support	Gradual relief as resources are replenished	Recomposition of roles and practices; enhanced use of existing expertise
Role of strategic actors	None (except reassuring stakeholders about safety)	Standard strategic planning and forecasting	Promoting and supporting innovative HR and skills management, in close connection with operations and work activities
Role of HR actors	Peripheral or reactive	Central, planning-oriented, macro-level	Enabling and integrative; closer to operational contexts
Role of operational actors	Compensate locally under pressure	Limited involvement, mainly as recipients of training	Active contributors to critical skill identification and development, as well as to the protection of critical tasks

(continued)

Table 11.1 (continued)

Analytical dimension	Adjustment	Restoration	Compensation
Role of safety actors	Locally patching and supporting operational actors	Usual duties, increased efforts to cope with production pressures	Active participation in novel skill management processes Enhancement/ mobilization of situated knowledge
Learning dynamics	Individual, informal, and invisible	Formalized, standardized, and often decontextualized	Collective, situated, and embedded in work activity
Main limitations	Produces hidden costs, safety degradation, and inequitable risk distribution	Slow, poorly suited to urgent needs, weak in specific or tacit skills	Organizationally demanding; requires coordination, trust, and innovation
Typical organizational outcome	Erosion of safety margins and performance over time	Partial recovery of capacity, often with temporal lag	Enhanced resilience under sustained constraints

References

C. Perrow, The limits of safety: the enhancements of a theory of accidents. J. Contingencies Crisis Manag. **2**(4) (1994)

P.R. Schulman, The negotiated order of organizational reliability. Adm. Soc. **25**(3), 353–372 (1993)

F. Taylor, *The Principles of Scientific Management* (Harper & Row, New York, 1911)

GPSR Compliance
The European Union's (EU) General Product Safety Regulation (GPSR) is a set of rules that requires consumer products to be safe and our obligations to ensure this.

If you have any concerns about our products, you can contact us on

ProductSafety@springernature.com

In case Publisher is established outside the EU, the EU authorized representative is:

Springer Nature Customer Service Center GmbH
Europaplatz 3
69115 Heidelberg, Germany

www.ingramcontent.com/pod-product-compliance
Ingram Content Group UK Ltd.
Pitfield, Milton Keynes, MK11 3LW, UK
UKHW022107180926
475695UK00001B/33
9783032229236